MESENCHYMAL STEM CELLS

MESENCHYMAL STEM CELLS

Biological Concepts, Current Advances, Opportunities, and Challenges

LEISHENG ZHANG
School of Medicine, Nankai University, Tianjin, China; Nanchang University, Nanchang, China

ZHONGCHAO HAN
State Key Laboratory of Experimental Hematology & National Clinical Research Center for Blood Disease, Institute of Hematology & Blood Diseases Hospital, Chinese Academy of Medical Sciences & Peking Union Medical Colleague, Tianjin, China

JIALUN WANG
National Engineering Technology Center for Cell Products & Jiangxi Research Center of Stem Cell Engineering, Jiangxi Health-Biotech Stem Cell Technology Co., Ltd., Shangrao, China

ZONGJIN LI
School of Medicine, Nankai University, Tianjin, China

QING HUANG
Key Laboratory of Radiation Technology and Biophysics, Institute of Biology and Hefei Institute of Physical Science, Chinese Academy of Sciences, Hefei, China

ELSEVIER

ACADEMIC PRESS
An imprint of Elsevier

Academic Press is an imprint of Elsevier
125 London Wall, London EC2Y 5AS, United Kingdom
525 B Street, Suite 1650, San Diego, CA 92101, United States
50 Hampshire Street, 5th Floor, Cambridge, MA 02139, United States
The Boulevard, Langford Lane, Kidlington, Oxford OX5 1GB, United Kingdom

ISBN 978-0-323-95346-7

For information on all Academic Press publications
visit our website at https://www.elsevier.com/books-and-journals

Publisher: Jonathan Simpson
Acquisitions Editor: Glyn Jones
Editorial Project Manager: Susan E. Ikeda
Production Project Manager: Fahmida Sultana
Cover Designer: Vicky Pearson Esser

Typeset by STRAIVE, India

Contents

Contributors

Cunrong Chen
Department of Critical Care Medicine, Union Hospital of Fujian Medical University, Fuzhou, China

Shuang Chen
Institute of Stem Cells, Health-Biotech (Tianjin) Stem Cell Research Institute Co., Ltd., Tianjin, China

Wenjing Du
The Postdoctoral Research Station, School of Medicine, Nankai University; National Engineering Research Center of Cell Products, AmCellGene Engineering Co., Ltd.; Tianjin Key Laboratory of Engineering Technologies for Cell Pharmaceutical, Tianjin, China

Zhikun Guo
Henan Key Laboratory of Medical Tissue Regeneration, Xinxiang Medical University, Xinxiang; Department of Cardiology, Zhengzhou Seventh People's Hospital, Zhengzhou, China

Zhongchao Han
Key Laboratory of Radiation Technology and Biophysics, Hefei Institute of Physical Science, Chinese Academy of Sciences, Hefei; National Engineering Technology Center for Cell Products & Jiangxi Research Center of Stem Cell Engineering, Jiangxi Health-Biotech Stem Cell Technology Co., Ltd., Shangrao; State Key Laboratory of Experimental Hematology & National Clinical Research Center for Blood Disease, Institute of Hematology & Blood Diseases Hospital, Chinese Academy of Medical Sciences & Peking Union Medical Colleague, Tianjin, China

Qianwen Hu
Division of Gastroenterology, The First Affiliated Hospital of Kunming Medical University, Kunming, China

Qing Huang
Key Laboratory of Radiation Technology and Biophysics, Hefei Institute of Physical Science, Chinese Academy of Sciences, Hefei, China

Pingping Jia
College of Medicine, Xinxiang University; Henan Key Laboratory of Medical Tissue Regeneration, Xinxiang Medical University, Xinxiang, China

Zongjin Li
School of Medicine; The Key Laboratory of Bioactive Materials, Ministry of Education, College of Life Sciences, Nankai University; Nankai University School of Medicine; Tianjin Key Laboratory of Human Development and Reproductive Regulation, Tianjin Central Hospital of Gynecology Obstetrics, Nankai University Affiliated Hospital of Obstetrics and Gynecology, Tianjin, China

Chao Liu
Key Laboratory of Radiation Technology and Biophysics, Hefei Institute of Physical Science, Chinese Academy of Sciences, Hefei, China

Yue Liu
School of Medicine; The Key Laboratory of Bioactive Materials, Ministry of Education, College of Life Sciences, Nankai University, Tianjin, China

Changsheng Shao
Key Laboratory of Radiation Technology and Biophysics, Hefei Institute of Physical Science, Chinese Academy of Sciences, Hefei, China

Baoquan Song
National Clinical Research Center for Hematologic Diseases, Jiangsu Institute of Hematology, The First Affiliated Hospital of Soochow University, Suzhou, China

Jialun Wang
National Engineering Technology Center for Cell Products & Jiangxi Research Center of Stem Cell Engineering, Jiangxi Health-Biotech Stem Cell Technology Co., Ltd., Shangrao, China

Hongju Yang
Division of Gastroenterology, The First Affiliated Hospital of Kunming Medical University, Kunming, China

Xingjie You
Division of Gastroenterology, The First Affiliated Hospital of Kunming Medical University, Kunming, China

Hao Yu
The Postdoctoral Research Station, School of Medicine, Nankai University; National Engineering Research Center of Cell Products, AmCellGene Engineering Co., Ltd.; Tianjin Key Laboratory of Engineering Technologies for Cell Pharmaceutical, Tianjin, China

Leisheng Zhang
Key Laboratory of Molecular Diagnostics and Precision Medicine for Surgical Oncology in Gansu Province & NHC Key Laboratory of Diagnosis and Therapy of Gastrointestinal Tumor, Gansu Provincial Hospital, Lanzhou; Key Laboratory of Radiation Technology and Biophysics, Hefei Institute of Physical Science, Chinese Academy of Sciences, Hefei; National Engineering Technology Center for Cell Products & Jiangxi Research Center of Stem Cell Engineering, Jiangxi Health-Biotech Stem Cell Technology Co., Ltd., Shangrao; School of Medicine, Nankai University; Tianjin Key Laboratory of Human Development and Reproductive Regulation, Tianjin Central Hospital of Gynecology Obstetrics, Nankai University Affiliated Hospital of Obstetrics and Gynecology; Institute of Stem Cells, Health-Biotech (Tianjin) Stem Cell Research Institute Co., Ltd.; State Key Laboratory of Experimental Hematology & National Clinical Research Center for Blood Disease, Institute of Hematology & Blood Diseases Hospital, Chinese Academy of Medical Sciences & Peking Union Medical Colleague, Tianjin, China

Wen Zhang
School of Medicine, Nankai University, Tianjin, China

Yuan-guang Zhao
Institute of Stem Cells, Health-Biotech (Tianjin) Stem Cell Research Institute Co., Ltd., Tianjin, China

Xinrun Zhou
School of Medicine, Nankai University, Tianjin, China

CHAPTER 1

Introduction of mesenchymal stem/stromal cells

Leisheng Zhang[a,b,c,d,e,*], Wenjing Du[f,g,h,*], Hao Yu[f,g,h,*], Xingjie You[i], Qianwen Hu[i], Hongju Yang[i], Shuang Chen[e], Yuan-guang Zhao[e], Zhongchao Han[c,d,j], and Jialun Wang[d]

[a]School of Medicine, Nankai University, Tianjin, China
[b]Key Laboratory of Molecular Diagnostics and Precision Medicine for Surgical Oncology in Gansu Province & NHC Key Laboratory of Diagnosis and Therapy of Gastrointestinal Tumor, Gansu Provincial Hospital, Lanzhou, China
[c]Key Laboratory of Radiation Technology and Biophysics, Hefei Institute of Physical Science, Chinese Academy of Sciences, Hefei, China
[d]National Engineering Technology Center for Cell Products & Jiangxi Research Center of Stem Cell Engineering, Jiangxi Health-Biotech Stem Cell Technology Co., Ltd., Shangrao, China
[e]Institute of Stem Cells, Health-Biotech (Tianjin) Stem Cell Research Institute Co., Ltd., Tianjin, China
[f]The Postdoctoral Research Station, School of Medicine, Nankai University, Tianjin, China
[g]National Engineering Research Center of Cell Products, AmCellGene Engineering Co., Ltd., Tianjin, China
[h]Tianjin Key Laboratory of Engineering Technologies for Cell Pharmaceutical, Tianjin, China
[i]Division of Gastroenterology, The First Affiliated Hospital of Kunming Medical University, Kunming, China
[j]State Key Laboratory of Experimental Hematology & National Clinical Research Center for Blood Disease, Institute of Hematology & Blood Diseases Hospital, Chinese Academy of Medical Sciences & Peking Union Medical Colleague, Tianjin, China

1.1 The definition of MSCs

Mesenchymal stem cells (MSCs), also known as mesenchymal stromal cells or medical signaling cells, are cell populations with self-renewal and multi-lineage differentiation capacity, which have been identified with unique hematologic-supporting and immunomodulatory properties [1–3]. MSCs were first isolated and identified from bone marrow in the 1960s, followed by other tissue sources, such as adult tissues (e.g., adipose tissue, dental pulp, synovial membrane) [3–8] and perinatal tissues (e.g., umbilical cord, placenta, amniotic membrane) [4,9–11].

For decades, MSCs were considered as plastic-adherent cells derived from the aforementioned sources. However, in 2005, The International Society for Cellular Therapy (ISCT) published the golden criteria for MSC clarification, including their fibroblast-like plastic-adherent morphology, high percentage of mesenchymal-associated biomarker expression (CD73, CD90, CD105), minimal expression of hematopoietic- or

*Co-first authors.

Mesenchymal Stem Cells
https://doi.org/10.1016/B978-0-323-95346-7.00003-8

immune-associated biomarkers (e.g., CD31, CD34, CD45, HLA-DR), and the multilineage differentiation potential towards adipocytes, osteoblasts, and chondrocytes [12].

1.2 The characteristics and biology of MSCs

MSCs with different origins reveal multifaceted similarities and variations in characteristics and biofunctions. Distinguished from pluripotent stem cells (PSCs), including embryonic stem cells (ESCs) and induced PSCs (iPSCs), MSCs are devoid of ethical conflicts on account of their sources including adult tissues, tissues surrounding the fetus, or medical waste after delivery [13]. Despite offering promising prospects for tissue engineering and regenerative medicine, bone marrow-derived mesenchymal stem cells (BM-MSCs) possess inherent defects for large-scale clinical application, such as low frequencies, limited proliferation, invasive sampling, and ethical risks [14].

Compared with adult tissue-derived MSCs, perinatal MSCs have superior self-renewal and differentiation properties, and in particular, their long-term in vitro proliferation further ensures the umbilical cord and placenta as an invaluable pool [4,9,13]. By comparison, Zhang and the colleagues found that human fetal bone marrow-derived MSCs (hF-MSCs) exhibited the most osteogenic capacity and the least immunogenicity, which further suggests that the anatomical and ontological origins of MSCs have a profound influence on biological functions and the resultant therapeutic effect. Interestingly, despite their comparable immunophenotypes, only hF-MSCs and human umbilical cord-derived MSCs (hUC-MSCs) were positive for the expression of pluripotency-related biomarkers OCT-4 and NANOG [14].

Longitudinal studies have indicated heterogeneity among MSCs with diverse origins due to anatomical structures in multifaceted biological phenotypes and biofunctions. For example, of the aforementioned MSCs, UC-MSCs and BM-MSCs reveal the most robust proliferation capacity and the widest range of application in clinical practice, respectively [1]. Of note, state-of-the-art studies have also indicated the variations in biofunction due to the origins and otherness of MSCs. For instance, Zhang et al. and Zhao et al., respectively, found that UC-MSCs at various passages showed diversity in the efficacy upon acute liver failure in rats and acute graft-versus-host disease (aGvHD) in mice, which was largely due to the alteration in phenotypic characteristics and molecular genetics [4,15].

In recent years, pioneering investigators have begun to explore the identification and preparation of unique MSC subsets with specific properties. For instance, Studle et al. identified the CD56^{+} subset, which has preferable bias towards chondrocytes and islet like cells, whereas Battula et al. found that the MSCA-1^{+} CD56^{+} subset revealed better pro-angiogenesis, immunomodulatory, and homing capacities compared with their corresponding counterparts, respectively [16,17]. Similarly, we recently found that the CD106^{+} (also known as VCAM-1^{+}) subset revealed superiority in pro-angiogenesis and immunomodulation over the CD106^{-} counterpart [2]. Moreover, with the aid of cytokine cocktail-based programming, we established a "one-step" procedure for high-efficiency CD106^{+} UC-MSCs preparation [3]. Compared with the CD106^{+} counterpart, CD106^{+} UC-MSCs revealed conservation in cytomorphology and immunophenotyping, combined with moderately enhanced pro-angiogenesis and bidirectional immunoregulatory properties [3]. Therefore, it will be interesting and attractive to further dissect specific subpopulations with unique biofunctions and prospects in the future, which will benefit MSC-based regenerative medicine.

1.3 The classification of MSCs

As mentioned above, MSCs can be divided into three catalogs, according to their origins, including adult tissues (e.g., adipose tissue, dental pulp), perinatal tissues (e.g., umbilical cord, placenta tissue, amniotic membrane), and derived from human pluripotent stem cells (e.g., hESCs, hiPSCs). In specific, MSCs derived from various adult tissues, including adipose, bone marrow, cartilage, tendon, and periosteum, have been reported. For example, Yao et al. reported the identification and comparison of apical papillary stem cells from human supernumerary teeth (SCAP-Ss) and dental pulp stem cells (DPSCs), and verified the preferable properties and efficacy of SACP-Ss on hepatic fibrosis in mice [7].

Perinatal tissues for MSC generation are also available, such as umbilical cord, placenta, amniotic membrane, amniotic fluid, and endometrium. For example, Xiang et al. and Zhao et al. reported the identification of UC-MSCs as well as their effectiveness upon early diabetic nephropathy and aGvHD, respectively [4,18]. Furthermore, Hou et al. and Du et al. verified the biological phenotypes and pro-angiogenic biofunction of placenta-derived MSCs (P-MSCs), respectively [2,9]. Very recently, Sun and colleagues determined the multifaceted biological and transcriptomic landscapes of MSCs derived from amniotic membrane (hAMSCs) [19].

State-of-the-art developments have also revealed the feasibility of MSC generation from human pluripotent stem cells (hPSCs), including human induced pluripotent stem cells (hiPSCs) and human embryonic stem cells (hESCs).

Simultaneously, we and our collaborators also dissected the properties and diversities of MSCs derived from healthy donors and patients with acute myelogenous leukemia (AML), acquired aplastic anemia, T-cell acute lymphoblastic leukemia (T-ALL), B-cell ALL (B-ALL), and type 2 diabetes mellitus (T2DM) [5,20–22].

1.4 The origins and development of MSCs

Current research indicates the origins of MSCs as being from various germ layers such as the mesoderm, trophoderm, endoderm, and ectoderm [23–29]. For instance, Lee and the colleagues reported the generation of neural crest stem cells (NCSCs) from hESCs by coculturing with MS-5 stromal cells, together with the further differentiation towards MSCs [27]. Wang et al. took advantage of the embryoid body (EB) model for TROP2$^+$ trophoblast cell generation and further maturation into MSCs from hESCs [30]. Furthermore, Kimbrel et al. derived hemangioblast from EB via a mesoderm intermediate stage [31]. Maxim and the colleagues reported the coculture of hESCs with OP9 stromal cells for mesoderm induction and the resultant MSCs and endothelial cell generation [25]. Taken together, MSCs have been derived from a variety of germ layers, which indicates the variations in the early development of MSCs with different origins.

For a long time, the regulatory mechanism underlying MSC generation and mesengenesis has remained largely obscure, which is largely attributable to the deficiency of procedures for high-efficiency MSC generation [32,33]. For this purpose, we and other investigators in the field have devoted effort to continuously optimizing the strategies, including coculture with stromal cells, the three-dimensional EB model, monolayer differentiation, and programming strategies. For instance, Vodyanik et al. cocultured hPSCs with OP-9 stromal cells, and demonstrated more efficient generation of mesenchymoangioblast during hPSC-MSCs induction. On the other hand, Ball and colleagues found the involvement of PDGFR inhibitor (PDGFR inhibitor-IV) was adequate for dedifferentiation of BM-MSCs into mesenchymoangioblast with mesoendoderm-associated biomarker expression [26].

In recent years, Deng et al. and Yu et al. demonstrated that more MSCs were induced by inhibition of IKK/NF-κB signaling and EZH2, respectively [24,33]. Meanwhile, a number of investigators demonstrated the facilitation of TGF-β signaling block by SB431542 addition during hESC-MSCs differentiation [34–37]. For example, Sánchez et al. added SB431542 into the hESCs culture medium for MSC induction, and a proportion of over 42% $CD73^{+}CD90^{+}$ MSC progenitor cells was induced within 28 days [38]. Of note, Wang et al. reported a novel procedure to rapidly and efficiently produce hESC-MSCs by modulating the BMP signaling-initiated trophoblast-like intermediate stage within 11–16 days [39].

However, most of the aforementioned procedures for hPSC-MSCs induction have inherent disadvantages, such as time-consuming, low-efficiency, scrapings and continuous passage for enrichment [23,32,38,40–43]. Recently, we and other investigators have taken advantage of the programming strategy for high-efficiency hPSC-MSC generation. For instance, Fukuta et al. combined SB431542 with CHIR99021 for neural crest cells (NCCs) generation and the resultant MSCs from hPSCs within 14 days [23]. In 2018, we reported high-efficiency hPSC-MSCs generation via a combination of ectopic MSX2 expression and cocktails (TGF-β, bFGF, CHIR99021, DAC) within 2 weeks [1]. Recently, Wei et al. reported the efficient generation of hESC-MSCs via OICR-9429 (a JAK signaling inhibitor) and decitabine (DAC) addition [44]. Similarly, we further took advantage of two JAK/STAT and BRD inhibitors (LLY-507, AZD5153), and verified the feasibility of the procedure for rapid and cost-effective hESC-MSCs generation within 14 days [45]. To date, considerable evidence has identified the viability of simultaneous programming via orchestrating epigenetic modification and signaling pathways for high-efficiency hPSC-MSC induction [44,45].

1.5 The safety and efficacy of MSCs

As mentioned above, MSCs possess self-renewal and multilineage differentiation potential. Consequently, MSCs offer promising prospects for exploring new therapeutic strategies for a variety of recurrent and refractory diseases and advancing MSC-based regenerative medicine. To illuminate the potential safety and efficacy of UC-MSCs at various stages, Zhao et al. conducted systematic characterization of the cellular and transcriptomic phenotypes, and confirmed the conservation in chromosome structure and nontumorigenicity, based on G-band analysis and subcutaneous tumor formation test, respectively [4].

In 2022, during an investigational new drug (IND) application in China, we put forward the importance and principles of quality control (QC) and therapeutic investigations of UC-MSCs for acute and chronic types of graft-versus-host disease (GvHD), which are the prerequisites to assure the reliability and consistency of the safety and quality of UC-MSCs before large-scale clinical practice and IND application [46]. Very recently, we further assessed the pharmacokinetics and confirmed the GMP-grade UC-MSCs product on a carbon tetrachloride (CCl4)-induced acute-on-chronic liver failure (ACLF) NOD-SCID mouse model and the concomitant therapeutic dosage for intravenous injection, which collectively provided overwhelming evidence for further investigating the dynamic metabolism and effectiveness in the future [47].

1.6 The regulation and molecular mechanism of MSCs

In recent years, numerous preclinical and clinical investigations have suggested the wide involvement of MSCs and their derivatives in the repair and functional reconstruction of various tissues and organ injuries via direct differentiation, trans-differentiation, paracrine and immunoregulation, and providing a hematopoietic microenvironment.

1.6.1 Secretion

Overall, secretion, including autocrine and paracrine, has been considered as a principal means in mediating the therapeutic effect of MSCs. During cell culture, exosomes and small extracellular vehicles (sEVs) are released into the culture medium and thus mediate the intercellular communications [44,48]. Actually, exosomes are nanovesicles with a particle size ranging from 20 nm to 200 nm, and play essential roles in a variety of physiological and pathological circumstances [49,50]. For instance, we recently reported the hESC-MSC-derived exosomes in facilitating pro-angiogenesis [44], together with the plasma-enriched exosomes in hindering megakaryocyte and platelet generation in patients with immune thrombocytopenia (ITP) [51].

Generally, the constituents of exosomes are largely attributable to origins including nucleic acids (e.g., mRNAs, miRNAs, ncRNAs, cirRNAs), proteins (e.g., cytokines), amino acids, bioactive lipids, metabolites, and subcellular fractions (e.g., mitochondria) [50,52,53]. To date, a variety of soluble cytokines have been identified in MSC-derived exosomes and sEVs, including transforming growth factor β (TGFβ), hepatocyte growth factor (HGF), prostaglandin-E2 (PGE2), and keratinocyte growth factor (KGF) [54,55]. For example, the MSC-released angiopoietin-1 and KGF in exosomes are involved in protecting the alveolar epithelium and endothelium from damage [56,57].

1.6.2 Differentiation

As referred to above, MSCs are multipotent stem cells with differentiation potential towards adiposes, osteoblasts, and chondrocytes, together with endothelial cells and neurocytes, which thus hold promising potential in tissue engineering and regenerative medicine. For example, autogenous and allogeneic MSCs have been proved to repair damaged lung tissues and degenerative orthopedic diseases via direct- or trans-differentiation into alveolar cells or osteoblasts to promote healing [58]. However, more and more studies seemingly have no place for the hypothesis of MSCs-derived functional cells in tissue reconstruction. For example, during rehabilitation of lung injury, there was no direct evidence of differentiation into lung tissues when considering the insufficiency of the effective retention rate in the body (<5%) [59,60].

1.7 The standardization and industrialization of MSCs

Longitudinal studies have indicated the successful outcomes of patients with refractory and recurrent diseases after local and systemic administration of MSCs. Unlike other types of cells for regenerative medicine, MSCs hold splendid advantages in dispensing with barriers in major histocompatibility complex (MHC), GvHD, and the resultant matching restrictions [61]. Of them, BM-MSCs have the most extensive range of applications in clinical practice, whereas UC-MSCs possess the most robust ex vivo proliferation capacity over other counterparts. Meanwhile, it is noteworthy that Zhang et al. and we recently reported the variations in the efficacy of UC-MSCs upon acute liver failure and acute GvHD (aGvHD), respectively [2,62].

Therefore, there is an urgent need to verify the feasibility of MSC-based cytothcrapy and the concomitant standardization of cell products via systematic and detailed study of the biological phenotype, biofunctions, and transcriptomic features. For example, Zhao and colleagues highlighted the conservation and multifaceted variations of UC-MSCs in cellular morphology, immunophenotyping, tri-lineage differentiation potential (adipocytes, osteoblasts, and chondrocytes), bidirectional immunoregulation, and hematopoietic-supporting effects [1].

Simultaneously, we and our colleagues have been dedicated to exploring potential subsets with unique features and preferable biofunctions to better satisfy the clinical demands. For example, Wei et al. reported the high-efficiency generation of $VCAM-1^+$ UC-MSCs with enhanced immunoregulatory and proangiogenic properties over the $VCAM-1^-$ counterpart by combining three cytokine cocktails [2]. Similarly, Studle et al. found that $MSCA-1^+CD56^+$ BM-MSCs revealed enhanced chondrogenesis, which suggested the subpopulation as a putative preferable cell source candidate for bone tissue engineering-based remedies [63]. Overall, considering the potential uncertainty and discrepancy as well as the large-scale application in clinical practice and new drug application (NDA), it is of great importance to verify the similarities and variations in industrial manufacturing and cell product standardization.

State-of-the-art studies have also indicated the necessity of launching public cell banks as well as the concomitant establishing of quality appraisal guidelines and good manufacturing practice (GMP); for example, the strategies for large-scale and cost-effective preparation of MSC products, including manual, automated, and semiautomated operations under pathogen-free conditions. For instance, Sun et al. recently conducted systematic comparison of the signatures of amniotic MSCs (AMSCs) under serum-containing or serum-free medium, which further indicated the viability of serum-free culture for the better preparation of standardized AMSC-based cell products for further drug development and clinical trials [64]. Additionally, as to the derivations such as exosomes and small extracellular vesicles (sEVs), it is more challenging to establish appropriate indicators and normalize the standardization for better adaptation to regenerative medicine in the future.

In the meantime, investigators and clinicians in the field have both noticed the importance of drug regimen for MSC-based cytotherapy. Up to December 26, 2022, a total of 1480 MSC-based trials have been registered according to the ClinicalTrials.gov website (https://www.clinicaltrials.gov).On the one hand, a variety of MSCs have been involved in clinical practice, including UC-MSCs, BM-MSCs, AMSCs, dental pulp stem cells (DPSCs), and adipose-derived MSCs (ADSCs). On the other hand, MSCs of different origins have been introduced for disease administration such as laryngotracheal stenosis, ulcerative colitis, bronchopulmonary dysplasia, type2 diabetes mellitus, knee osteoarthritis, lumbar disc degeneration disease, acute respiratory distress syndrome (ARDS), and acute-on-chronic liver failure (ACLF). On the basis of current rationale and clinical trials, we are greatly encouraged by the inspiring progress of MSC-based cytotherapy, and we also noticed the otherness and even contradictoriness in outcomes, which collectively highlights the importance of the principles of MSC-based remedies in future.

Acknowledgment

This work was supported by grants from the National Natural Science Foundation of China (82260031, 31900460), supported by Jiangxi Provincial Natural Science Foundation (20224BAB206077, 20212BAB216073), the project Youth Fund, supported by Shandong Provincial Natural Science Foundation (ZR2020QC097), Gansu Provincial Hospital Intra-Hospital Research Fund Project (22GSSYB-6), The 2022 Master/Doctor/Postdoctoral program of NHC Key Laboratory of Diagnosis and Therapy of Gastrointestinal Tumor (NHCDP2022004, NHCDP2022008), the Nonprofit Central Research Institute Fund of Chinese Academy of Medical Sciences (2019PT320005), and the 2021 Central-Guided Local Science and Technology Development Fund (ZYYDDFFZZJ-1), Key project funded by Department of Science and Technology of Shangrao City (2020G002, 2020AB002, 2020K003, 2021F013, 2022AB003), Double Thousand Plan of Jiangxi Province (2022 to LSZ).

Conflict of interest

The coauthors declare that there are no competing interests and all the coauthors consent to publish the data.

References

[1] Zhang L, Wang H, Liu C, Wu Q, Su P, Wu D, Guo J, Zhou W, Xu Y, Shi L, et al. MSX2 initiates and accelerates mesenchymal stem/stromal cell specification of hPSCs by regulating TWIST1 and PRAME. Stem Cell Rep 2018;11(2):497–513.

[2] Du W, Li X, Chi Y, Ma F, Li Z, Yang S, Song B, Cui J, Ma T, Li J, et al. VCAM-1+ placenta chorionic villi-derived mesenchymal stem cells display potent pro-angiogenic activity. Stem Cell Res Ther 2016;7:49.

[3] Wei Y, Zhang L, Chi Y, Ren X, Gao Y, Song B, Li C, Han Z, Zhang L, Han Z. High-efficient generation of VCAM-1(+) mesenchymal stem cells with multidimensional superiorities in signatures and efficacy on aplastic anaemia mice. Cell Prolif 2020;53 (8), e12862.

[4] Zhao Q, Zhang L, Wei Y, Yu H, Zou L, Huo J, Yang H, Song B, Wei T, Wu D, et al. Systematic comparison of hUC-MSCs at various passages reveals the variations of signatures and therapeutic effect on acute graft-versus-host disease. Stem Cell Res Ther 2019;10(1):354.

[5] Wang L, Zhang L, Liang X, Zou J, Liu N, Liu T, Wang G, Ding X, Liu Y, Zhang B, et al. Adipose tissue-derived stem cells from type 2 diabetics reveal conservative alterations in multidimensional characteristics. Int J Stem Cells 2020;13(2):268–78.

[6] Huo J, Zhang L, Ren X, Li C, Li X, Dong P, Zheng X, Huang J, Shao Y, Ge M, et al. Multifaceted characterization of the signatures and efficacy of mesenchymal stem/stromal cells in acquired aplastic anemia. Stem Cell Res Ther 2020;11(1):59.

[7] Yao J, Chen N, Wang X, Zhang L, Huo J, Chi Y, Li Z, Han Z. Human supernumerary teeth-derived apical papillary stem cells possess preferable characteristics and efficacy on hepatic fibrosis in mice. Stem Cells Int 2020;2020:6489396.

[8] Gadkari R, Zhao L, Teklemariam T, Hantash BM. Human embryonic stem cell derived-mesenchymal stem cells: an alternative mesenchymal stem cell source for regenerative medicine therapy. Regen Med 2014;9(4):453–65.

[9] Hou H, Zhang L, Duan L, Liu Y, Han Z, Li Z, Cao X. Spatio-temporal metabolokinetics and efficacy of human placenta-derived mesenchymal stem/stromal cells on mice with refractory Crohn's-like enterocutaneous fistula. Stem Cell Rev Rep 2020;16 (6):1292–304.

[10] Rosada C, Justesen J, Melsvik D, Ebbesen P, Kassem M. The human umbilical cord blood: a potential source for osteoblast progenitor cells. Calcif Tissue Int 2003;72 (2):135–42.

[11] Javed MJ, Mead LE, Prater D, Bessler WK, Foster D, Case J, Goebel WS, Yoder MC, Haneline LS, Ingram DA. Endothelial colony forming cells and mesenchymal stem cells are enriched at different gestational ages in human umbilical cord blood. Pediatr Res 2008;64(1):68–73.

[12] Horwitz EM, Le Blanc K, Dominici M, Mueller I, Slaper Cortenbach I, Marini FC, Deans RJ, Krause DS, Keating A. International Society for Cellular T: clarification of the nomenclature for MSC: the International Society for Cellular Therapy position statement. Cytotherapy 2005;7(5):393–5.

[13] Kulus M, Sibiak R, Stefanska K, Zdun M, Wieczorkiewicz M, Piotrowska-Kempisty H, Jaskowski JM, Bukowska D, Ratajczak K, Zabel M, et al. Mesenchymal stem/stromal cells derived from human and animal perinatal tissues-origins, characteristics, signaling pathways, and clinical trials. Cell 2021;10(12):3278.

[14] Zhang ZY, Teoh SH, Chong MS, Schantz JT, Fisk NM, Choolani MA, Chan J. Superior osteogenic capacity for bone tissue engineering of fetal compared with perinatal and adult mesenchymal stem cells. Stem Cells 2009;27(1):126–37.

[15] Zhang Y, Li Y, Li W, Cai J, Yue M, Jiang L, Xu R, Zhang L, Li J, Zhu C. Therapeutic effect of human umbilical cord mesenchymal stem cells at various passages on acute liver failure in rats. Stem Cells Int 2018;2018:7159465.

[16] Studle C, Occhetta P, Geier F, Mehrkens A, Barbero A, Martin I. Challenges toward the identification of predictive markers for human mesenchymal stromal cells chondrogenic potential. Stem Cells Transl Med 2019;8(2):194–204.

[17] Battula VL, Treml S, Bareiss PM, Gieseke F, Roelofs H, de Zwart P, Muller I, Schewe B, Skutella T, Fibbe WE, et al. Isolation of functionally distinct mesenchymal stem cell subsets using antibodies against CD56, CD271, and mesenchymal stem cell antigen-1. Haematologica 2009;94(2):173–84.

[18] Xiang E, Han B, Zhang Q, Rao W, Wang Z, Chang C, Zhang Y, Tu C, Li C, Wu D. Human umbilical cord-derived mesenchymal stem cells prevent the progression of early diabetic nephropathy through inhibiting inflammation and fibrosis. Stem Cell Res Ther 2020;11(1):336.

[19] Sun Y, Wang TE, Hu Q, Zhang W, Zeng Y, Lai X, Zhang L, Shi M. Systematic comparation of the biological and transcriptomic landscapes of human amniotic mesenchymal stem cells under serum-containing and serum-free conditions. Stem Cell Res Ther 2022;13(1):490.

[20] Zhang L, Chi Y, Wei Y, Zhang W, Wang F, Zhang L, Zou L, Song B, Zhao X, Han Z. Bone marrow-derived mesenchymal stem/stromal cells in patients with acute myeloid leukemia reveal transcriptome alterations and deficiency in cellular vitality. Stem Cell Res Ther 2021;12(1):365.

[21] Wang J, Liu X, Qiu Y, Shi Y, Cai J, Wang B, Wei X, Ke Q, Sui X, Wang Y, et al. Cell adhesion-mediated mitochondria transfer contributes to mesenchymal stem

cell-induced chemoresistance on T cell acute lymphoblastic leukemia cells. J Hematol Oncol 2018;11(1):11.
[22] Chen F, Zhong X, Dai Q, Li K, Zhang W, Wang J, Zhao Y, Shen J, Xiao Z, Xing H, et al. Human umbilical cord MSC delivered-soluble TRAIL inhibits the proliferation and promotes apoptosis of B-ALL cell in vitro and in vivo. Pharmaceuticals (Basel) 2022;15(11):1391.
[23] Fukuta M, Nakai Y, Kirino K, Nakagawa M, Sekiguchi K, Nagata S, Matsumoto Y, Yamamoto T, Umeda K, Heike T, et al. Derivation of mesenchymal stromal cells from pluripotent stem cells through a neural crest lineage using small molecule compounds with defined media. PLoS One 2014;9(12), e112291.
[24] Yu Y, Deng P, Yu B, Szymanski JM, Aghaloo T, Hong C, Wang CY. Inhibition of EZH2 promotes human embryonic stem cell differentiation into mesoderm by reducing H3K27me3. Stem Cell Rep 2017;9(3):752–61.
[25] Vodyanik MA, Yu J, Zhang X, Tian S, Stewart R, Thomson JA, Slukvin II. A mesoderm-derived precursor for mesenchymal stem and endothelial cells. Cell Stem Cell 2010;7(6):718–29.
[26] Ball SG, Worthington JJ, Canfield AE, Merry CL, Kielty CM. Mesenchymal stromal cells: inhibiting PDGF receptors or depleting fibronectin induces mesodermal progenitors with endothelial potential. Stem Cells 2014;32(3):694–705.
[27] Lee G, Kim H, Elkabetz Y, Al Shamy G, Panagiotakos G, Barberi T, Tabar V, Studer L. Isolation and directed differentiation of neural crest stem cells derived from human embryonic stem cells. Nat Biotechnol 2007;25(12):1468–75.
[28] Morikawa S, Mabuchi Y, Niibe K, Suzuki S, Nagoshi N, Sunabori T, Shimmura S, Nagai Y, Nakagawa T, Okano H, et al. Development of mesenchymal stem cells partially originate from the neural crest. Biochem Biophys Res Commun 2009;379 (4):1114–9.
[29] Slukvin II, Vodyanik M. Endothelial origin of mesenchymal stem cells. Cell Cycle 2011;10(9):1370–3.
[30] Barberi T, Willis LM, Socci ND, Studer L. Derivation of multipotent mesenchymal precursors from human embryonic stem cells. PLoS Med 2005;2(6), e161.
[31] Kimbrel EA, Kouris NA, Yavanian GJ, Chu J, Qin Y, Chan A, Singh RP, McCurdy D, Gordon L, Levinson RD, et al. Mesenchymal stem cell population derived from human pluripotent stem cells displays potent immunomodulatory and therapeutic properties. Stem Cells Dev 2014;23(14):1611–24.
[32] Gibson JD, O'Sullivan MB, Alaee F, Paglia DN, Yoshida R, Guzzo RM, Drissi H. Regeneration of articular cartilage by human ESC-derived mesenchymal progenitors treated sequentially with BMP-2 and Wnt5a. Stem Cells Transl Med 2017;6 (1):40–50.
[33] Deng P, Zhou C, Alvarez R, Hong C, Wang CY. Inhibition of IKK/NF-kappaB signaling enhances differentiation of mesenchymal stromal cells from human embryonic stem cells. Stem Cell Rep 2016;6(4):456–65.
[34] Tran NT, Trinh QM, Lee GM, Han YM. Efficient differentiation of human pluripotent stem cells into mesenchymal stem cells by modulating intracellular signaling pathways in a feeder/serum-free system. Stem Cells Dev 2012;21(7):1165–75.
[35] Mahmood A, Harkness L, Schroder HD, Abdallah BM, Kassem M. Enhanced differentiation of human embryonic stem cells to mesenchymal progenitors by inhibition of TGF-beta/activin/nodal signaling using SB-431542. J Bone Miner Res 2010;25 (6):1216–33.
[36] Whitworth DJ, Frith JE, Frith TJ, Ovchinnikov DA, Cooper-White JJ, Wolvetang EJ. Derivation of mesenchymal stromal cells from canine induced pluripotent stem cells by inhibition of the TGFbeta/activin signaling pathway. Stem Cells Dev 2014;23 (24):3021–33.

[37] Denis JA, Rochon-Beaucourt C, Champon B, Pietu G. Global transcriptional profiling of neural and mesenchymal progenitors derived from human embryonic stem cells reveals alternative developmental signaling pathways. Stem Cells Dev 2011;20(8):1395–409.
[38] Sanchez L, Gutierrez-Aranda I, Ligero G, Rubio R, Munoz-Lopez M, Garcia-Perez JL, Ramos V, Real PJ, Bueno C, Rodriguez R, et al. Enrichment of human ESC-derived multipotent mesenchymal stem cells with immunosuppressive and anti-inflammatory properties capable to protect against experimental inflammatory bowel disease. Stem Cells 2011;29(2):251–62.
[39] Wang X, Lazorchak AS, Song L, Li E, Zhang Z, Jiang B, Xu RH. Immune modulatory mesenchymal stem cells derived from human embryonic stem cells through a trophoblast-like stage. Stem Cells 2016;34(2):380–91.
[40] Olivier EN, Rybicki AC, Bouhassira EE. Differentiation of human embryonic stem cells into bipotent mesenchymal stem cells. Stem Cells 2006;24(8):1914–22.
[41] Boyd NL, Robbins KR, Dhara SK, West FD, Stice SL. Human embryonic stem cell-derived mesoderm-like epithelium transitions to mesenchymal progenitor cells. Tissue Eng A 2009;15(8):1897–907.
[42] Lian Q, Lye E, Suan Yeo K, Khia Way Tan E, Salto Tellez M, Liu TM, Palanisamy N, El Oakley RM, Lee EH, Lim B, et al. Derivation of clinically compliant MSCs from CD105+, CD24- differentiated human ESCs. Stem Cells 2007;25(2):425–36.
[43] Kopher RA, Penchev VR, Islam MS, Hill KL, Khosla S, Kaufman DS. Human embryonic stem cell-derived CD34+ cells function as MSC progenitor cells. Bone 2010;47(4):718–28.
[44] Wei Y, Hou H, Zhang L, Zhao N, Li C, Huo J, Liu Y, Zhang W, Li Z, Liu D, et al. JNKi- and DAC-programmed mesenchymal stem/stromal cells from hESCs facilitate hematopoiesis and alleviate hind limb ischemia. Stem Cell Res Ther 2019;10(1):186.
[45] Zhang L, Wei Y, Chi Y, Liu D, Yang S, Han Z, Li Z. Two-step generation of mesenchymal stem/stromal cells from human pluripotent stem cells with reinforced efficacy upon osteoarthritis rabbits by HA hydrogel. Cell Biosci 2021;11(1):6.
[46] Wang A, Zhang L, Zhao M, Yu H. Quality control and therapeutic investigations of mesenchymal stem/stromal cells during investigational new drug application for GvHD administration in China. Curr Stem Cell Res Ther 2022.
[47] Yu H, Feng Y, Du W, Zhao M, Jia H, Wei Z, Yan S, Han Z, Zhang L, Li Z, et al. Off-the-shelf GMP-grade UC-MSCs as therapeutic drugs for the amelioration of CCl4-induced acute-on-chronic liver failure in NOD-SCID mice. Int Immunopharmacol 2022;113(Pt A), 109408.
[48] Khatri M, Richardson LA, Meulia T. Mesenchymal stem cell-derived extracellular vesicles attenuate influenza virus-induced acute lung injury in a pig model. Stem Cell Res Ther 2018;9(1):17.
[49] Ha DH, Kim HK, Lee J, Kwon HH, Park GH, Yang SH, Jung JY, Choi H, Lee JH, Sung S, et al. Mesenchymal stem/stromal cell-derived exosomes for immunomodulatory therapeutics and skin regeneration. Cell 2020;9(5):1157.
[50] Kalluri R, LeBleu VS. The biology, function, and biomedical applications of exosomes. Science 2020;367(6478), eaau6977.
[51] Miao W, Song B, Shi B, Wan Q, Lv Q, Chen H, Zhu M, Zhang L, Han Y, Wu D. Immune thrombocytopenia plasma-derived exosomes impaired megakaryocyte and platelet production through an apoptosis pathway. Thromb Haemost 2020;121:495–505.
[52] Baglio SR, Rooijers K, Koppers-Lalic D, Verweij FJ, Perez Lanzon M, Zini N, Naaijkens B, Perut F, Niessen HW, Baldini N, et al. Human bone marrow- and adipose-mesenchymal stem cells secrete exosomes enriched in distinctive miRNA and tRNA species. Stem Cell Res Ther 2015;6:127.

[53] Eirin A, Zhu XY, Puranik AS, Woollard JR, Tang H, Dasari S, Lerman A, van Wijnen AJ, Lerman LO. Integrated transcriptomic and proteomic analysis of the molecular cargo of extracellular vesicles derived from porcine adipose tissue-derived mesenchymal stem cells. PLoS One 2017;12(3), e0174303.

[54] Lee JW, Fang X, Gupta N, Serikov V, Matthay MA. Allogeneic human mesenchymal stem cells for treatment of *E. coli* endotoxin-induced acute lung injury in the ex vivo perfused human lung. Proc Natl Acad Sci U S A 2009;106(38):16357–62.

[55] Zhu YG, Feng XM, Abbott J, Fang XH, Hao Q, Monsel A, Qu JM, Matthay MA, Lee JW. Human mesenchymal stem cell microvesicles for treatment of Escherichia coli endotoxin-induced acute lung injury in mice. Stem Cells 2014;32(1):116–25.

[56] Lee JW, Fang X, Krasnodembskaya A, Howard JP, Matthay MA. Concise review: mesenchymal stem cells for acute lung injury: role of paracrine soluble factors. Stem Cells 2011;29(6):913–9.

[57] Tang XD, Shi L, Monsel A, Li XY, Zhu HL, Zhu YG, Qu JM. Mesenchymal stem cell microvesicles attenuate acute lung injury in mice partly mediated by Ang-1 mRNA. Stem Cells 2017;35(7):1849–59.

[58] Behnke J, Kremer S, Shahzad T, Chao CM, Bottcher-Friebertshauser E, Morty RE, Bellusci S, Ehrhardt H. MSC based therapies-new perspectives for the injured lung. J Clin Med 2020;9(3):682.

[59] Krause DS, Theise ND, Collector MI, Henegariu O, Hwang S, Gardner R, Neutzel S, Sharkis SJ. Multi-organ, multi-lineage engraftment by a single bone marrow-derived stem cell. Cell 2001;105(3):369–77.

[60] Zhang LS, Yu Y, Yu H, Han ZC. Therapeutic prospects of mesenchymal stem/stromal cells in COVID-19 associated pulmonary diseases: from bench to bedside. World J Stem Cells 2021;13(8):1058–71.

[61] Xu ZL, Huang XJ. Optimizing allogeneic grafts in hematopoietic stem cell transplantation. Stem Cells Transl Med 2021;10(Suppl 2):S41–7.

[62] Zhang Y, Li Y, Li W, Cai J, Yue M, Jiang L, Xu R, Zhang L, Li J, Zhu C. Therapeutic effect of human umbilical cord mesenchymal stem cells at various passages on acute liver failure in rats. Stem Cells Int 2018;2018:7159465.

[63] Stüdle C, Occhetta P, Geier F, Mehrkens A, Barbero A, Martin I. Challenges toward the identification of predictive markers for human mesenchymal stromal cells chondrogenic potential. Stem Cells Transl Med 2019;8(2):194–204.

[64] Sun Y, Wang TE, Hu Q, Zhang W, Zeng Y, Lai X, Zhang L, Shi M. Systematic comparation of the biological and transcriptomic landscapes of human amniotic mesenchymal stem cells under serum-containing and serum-free conditions. Stem Cell Res Ther 2022;13(1):490.

CHAPTER 2

The application of MSCs in regenerative medicine

Leisheng Zhang[a,b,c,d,e], Baoquan Song[f], Cunrong Chen[g], Yuan-guang Zhao[e], Shuang Chen[e], Zhongchao Han[c,d,h], and Jialun Wang[d]

[a]School of Medicine, Nankai University, Tianjin, China
[b]Key Laboratory of Molecular Diagnostics and Precision Medicine for Surgical Oncology in Gansu Province & NHC Key Laboratory of Diagnosis and Therapy of Gastrointestinal Tumor, Gansu Provincial Hospital, Lanzhou, China
[c]Key Laboratory of Radiation Technology and Biophysics, Hefei Institute of Physical Science, Chinese Academy of Sciences, Hefei, China
[d]National Engineering Technology Center for Cell Products & Jiangxi Research Center of Stem Cell Engineering, Jiangxi Health-Biotech Stem Cell Technology Co., Ltd., Shangrao, China
[e]Institute of Stem Cells, Health-Biotech (Tianjin) Stem Cell Research Institute Co., Ltd., Tianjin, China
[f]National Clinical Research Center for Hematologic Diseases, Jiangsu Institute of Hematology, The First Affiliated Hospital of Soochow University, Suzhou, China
[g]Department of Critical Care Medicine, Union Hospital of Fujian Medical University, Fuzhou, China
[h]State Key Laboratory of Experimental Hematology & National Clinical Research Center for Blood Disease, Institute of Hematology & Blood Diseases Hospital, Chinese Academy of Medical Sciences & Peking Union Medical Colleague, Tianjin, China

2.1 The application of MSCs in hematological diseases

Hematopoietic stem cell transplantation (HSCT) has been widely used for treatment of various hematological diseases, yet is still seriously limited by its sources and the concomitant graft-versus-host disease (GvHD) [1,2]. In recent years, due to the unique immunoregulatory properties of MSCs, cotransplantation of hematopoietic stem cells (HSCs) and MSCs can effectively reduce the adverse reactions, and thus, effectively improve the efficiency of hematopoietic reconstruction and the survival rate of patients [3]. For example, Wu and colleagues enrolled 50 patients with recurrent and refractory hematologic malignancies, and those after cotransplantation of HSCs and MSCs revealed 1-year and 2-year disease-free/progression-free survival rates of 77.8% and 66%, respectively [3]. In 2013, Li and colleagues reported the cotransplantation of MSCs and $CD34^+$ HSCs into NSG mice, and found that both hESC-MSCs and BM-MSCs revealed hematopoietic-supporting capacity [4] (Table 1).

Mesenchymal Stem Cells
https://doi.org/10.1016/B978-0-323-95346-7.00005-1

Table 1 Representative MSC-based trials for hematological disease treatment.

Disease	NCT number	Status	Phases	Enrollment
GvHD, MS	NCT00081055	Withdrawn	Phase 2	0
GvHD	NCT00136903	Completed	Phase 2	32
GvHD	NCT00314483	Unknown status	Phase 1,2	25
GvHD	NCT00366145	Completed	Phase 3	260
GvHD	NCT00447460	Unknown status	Phase 1,2	15
GvHD, MS	NCT00498316	Completed	Phase 1	98
GvHD	NCT00972660	Unknown status	Phase 2	52
CML	NCT01045382	Terminated	Phase 2	39
aGvHD	NCT01765634	Unknown status	Phase 2	40
cGvHD	NCT01765660	Unknown status	Phase 2	60
cGvHD	NCT01956903	Completed	Phase 1,2	15
ALL, AML, MS, CLL	NCT02181478	Completed	Early Phase 1	6
ALL, AML, MS	NCT02181478	Completed	Early Phase 1	6
GvHD	NCT02824653	Completed	Phase 1,2	10
CML, MS	NCT03096782	Completed	Phase 2	6
AML	NCT03233074	Recruiting		150
AML	NCT03918655	Active	186	
GvHD	NCT05152160	Recruiting	Phase 1,2	10
MM	NCT05113342	Recruiting	Phase 1,2	20
CLL, MS	NCT01624701	Terminated	Phase 1,2	3

2.2 The application of MSCs in digestive diseases

In recent years, many reports have indicated the application of MSCs in the treatment of digestive diseases such as hepatic diseases (e.g., liver fibrosis, cirrhosis, end-stage liver disease, acute-on-chronic liver failure) and gastrointestinal diseases (e.g., ulcerative colitis, acute colitis, Crohn's disease) [5–7]. As to hepatic diseases, Ghavamzadeh et al. and Yao et al. verified the efficacy of MSC-based cytotherapy for hepatic iron overload and liver fibrosis via modulating the secretion of proinflammatory factors, respectively [7,8]. Of note, Terai and colleagues reported the efficient remission of nine patients with liver cirrhosis by systematic infusion of autologous BM-MSCs [9,10].

As to gastrointestinal diseases, Zhang et al. took advantage of the allogeneic UC-MSCs infusion, and found that most of the enrolled patients manifested a sharp decline in Crohn's Disease Activity Index (CDAI), corticosteroid intake, and Harvey-Bradshaw Index (HBI). In 2011, Sanchez and colleagues reported the efficient remission of mice with experimental inflammatory bowel disease via systematic infusion of hPSC-MSCs [11]. In 2016, Wang et al. verified that hPSC-MSCs and BM-MSCs exhibited comparable efficacy upon DSS-induced acute colitis mice [12]. Additionally, investigators in the field further verified the preferable outcomes of mice with TNBS- or DSS-induced inflammatory bowel disease (IBD) by employing the MSCs/hydrogel composite, and in particular, the efficacy upon refractory fistulas and inflammatory damage [6,13,14]. In 2020, we took advantage of the Crohn's disease (CD) model and demonstrated the spatiotemporal metabolokinetics and efficacy of the dual-fluorescence expressing P-MSCs on intractable CD with enterocutaneous fistula (ECF). Our findings indicated the therapeutic mechanism of P-MSCs via simultaneously downregulating reactive oxygen species (ROS) and accelerating neovascularization [5] (Table 2).

Table 2 Representative MSC-based trials for digestive disease treatment.

Disease	NCT number	Status	Phases	Enrollment
Crohn's Disease	NCT00294112	Completed	Phase 2	10
Liver Failure, Cirrhosis	NCT00420134	Completed	Phase 1, 2	30
Cirrhosis	NCT00476060	Unknown status	Phase 2	36
Crohn's Disease	NCT00482092	Completed	Phase 3	330
Crohn's Disease	NCT00543374	Completed	Phase 3	98
Liver Cirrhosis	NCT00913289	Terminated	Phase 1	6
Liver Cirrhosis	NCT00976287	Unknown status	Phase 2	50
Liver Cirrhosis	NCT00993941	Unknown status	Phase 2	60
Liver Cirrhosis	NCT01062750	Completed	Not Applicable	4
Crohn's Disease	NCT01144962	Completed	Phase 1, 2	21
Crohn's Disease	NCT01157650	Completed	Phase 1, 2	15
Liver Cirrhosis	NCT01220492	Completed	Phase 1, 2	266

Continued

Table 2 Representative MSC-based trials for digestive disease treatment—cont'd

Disease	NCT number	Status	Phases	Enrollment
Ulcerative Colitis	NCT01221428	Unknown status	Phase 1, 2	50
Liver Cirrhosis	NCT01223664	Unknown status	Phase 2	60
Liver Cirrhosis	NCT01224327	Unknown status	Phase 1, 2	50
Liver Cirrhosis	NCT01233102	Suspended	Phase 1, 2	200
Crohn's Disease	NCT01233960	Completed	Phase 3	73
Perianal Fistula	NCT01314092	Terminated	Phase 2	15
Liver Cirrhosis	NCT01342250	Completed	Phase 1, 2	20
Wilson's Disease	NCT01378182	Completed	Not Applicable	10
Liver Failure	NCT01429038	Completed	Phase 1, 2	40
Primary Biliary Cirrhosis	NCT01440309	Unknown status	Phase 1	20
Liver Fibrosis	NCT01454336	Completed	Phase 1	3
Liver Cirrhosis	NCT01483248	Unknown status	Phase 1, 2	50
Liver Cirrhosis	NCT01499459	Unknown status	Not Applicable	25
Crohn's Disease	NCT01510431	No longer available		
Crohn's Disease	NCT01540292	Terminated	Phase 1, 2	13
Crohn's Disease	NCT01541579	Completed	Phase 3	278
Liver Cirrhosis	NCT01573923	Unknown status	Phase 1, 2	320
Alcoholic Liver Cirrhosis	NCT01591200	Completed	Phase 2	40
Perianal Fistula	NCT01623453	Terminated		6
Primary Biliary Cirrhosis	NCT01662973	Unknown status	Phase 1, 2	100
Liver Failure	NCT01724398	Unknown status	Phase 1, 2	120
End Stage Liver Disease	NCT01728727	Unknown status	Phase 1, 2	240
Alcoholic Liver Cirrhosis	NCT01741090	Unknown status	Phase 2	12
Crohn's Disease	NCT01874015	Unknown status	Phase 1	10
Alcoholic Liver Cirrhosis	NCT01875081	Completed	Phase 2	72
Liver Cirrhosis	NCT01877759	Unknown status	Phase 1, 2	20
Ulcerative Colitis	NCT01914887	Unknown status	Phase 1, 2	8

Table 2 Representative MSC-based trials for digestive disease treatment—cont'd

Disease	NCT number	Status	Phases	Enrollment
Perianal Crohn's Disease	NCT01915927	Completed	Phase 1	20
Crohn's Disease	NCT02000362	Unknown status	Phase 1, 2	24
Crohn's Disease	NCT02403232	Unknown status	Phase 2	10
Ulcerative Colitis	NCT02442037	Unknown status	Phase 1, 2	30
Crohn's Disease	NCT02445547	Completed	Phase 1, 2	82
Crohn's Disease	NCT02580617	Recruiting	Phase 1	9
Hepatic Cirrhosis	NCT02652351	Unknown status	Phase 1	20
Crohn's Disease	NCT02677350	Withdrawn	Phase 1	0
Liver Cirrhosis	NCT02705742	Unknown status	Phase 1, 2	5
Decompensated Cirrhosis	NCT02786017	Unknown status	Phase 1, 2	40
Liver Failure	NCT02812121	Unknown status	Phase 2	261
ACLF	NCT02857010	Unknown status	Phase 1	30
Crohn's Disease	NCT02926300	Unknown status		24
Liver Cirrhosis	NCT02943889	Unknown status	Phase 1, 2	40
IBD	NCT03115749	Terminated	Not Applicable	2
Crohn's Disease	NCT03183661	Unknown status		9
Liver Cirrhosis	NCT03209986	Unknown status	Not Applicable	200
Crohn's Disease	NCT03220243	Completed	Phase 1	5
Decompensated Liver Cirrhosis	NCT03254758	Recruiting	Phase 1, 2	27
IBD	NCT03299413	Unknown status	Phase 1, 2	20
Crohn's Disease	NCT03449069	Recruiting	Phase 1	5
Renal Cirrhosis	NCT03460223	Unknown status	Phase 1, 2	30
Liver Cirrhosis	NCT03460795	Not recruiting	Phase 1, 2	30
Decompensated Liver Cirrhosis	NCT03472742	Recruiting		21

Continued

Table 2 Representative MSC-based trials for digestive disease treatment—cont'd

Disease	NCT number	Status	Phases	Enrollment
Decompensated Liver Cirrhosis	NCT03529136	Unknown status	Phase 2	252
Ulcerative Colitis (UC)	NCT03609905	Unknown status	Phase 1, 2	50
Liver Cirrhosis	NCT03626090	Recruiting	Phase 1, 2	20
Acute Liver Failure, ACLF	NCT03629015	Withdrawn	Phase 1	0
Primary Biliary Cirrhosis	NCT03668145	Unknown status	Not Applicable	140
ACLF	NCT03668171	Unknown status	Not Applicable	200
Hepatitis B	NCT03826433	Recruiting	Phase 1	20
Alcoholic Liver Cirrhosis	NCT03838250	Unknown status	Phase 1	10
ACLF	NCT03860155	Terminated	Phase 1, 2	5
ACLF	NCT03863002	Unknown status	Phase 1, 2	45
Decompensated Liver Cirrhosis	NCT03945487	Recruiting	Phase 2	200
Crohn's Disease	NCT04073472	Withdrawn	Phase 1	0
Liver Cirrhosis	NCT04243681	Completed	Phase 4	5
Ulcerative Colitis	NCT04312113	Recruiting	Phase 1	20
Liver Cirrhosis	NCT04357600	Recruiting	Phase 1, 2	12
Perianal Crohn's disease	NCT04519671	Recruiting	Phase 1, 2	40
Crohn's Disease	NCT04519684	Recruiting	Phase 1, 2	40
Crohn's Disease	NCT04519697	Recruiting	Phase 1, 2	40
Primary Biliary Cirrhosis	NCT04522869	Unknown status	Phase 1, 2	34
Ulcerative Colitis	NCT04543994	Recruiting	Phase 1, 2	24
Crohn's Colitis	NCT04548583	Recruiting	Phase 1, 2	24
Alcoholic Cirrhosis	NCT04689152	Recruiting	Phase 3	200
Anal Fistula	NCT04750499	Recruiting	Not Applicable	12
Crohn's Disease	NCT04791878	Recruiting	Phase 1	10
ACLF	NCT04822922	Not recruiting	Phase 2	16
Crohn's Disease	NCT04847739	Recruiting	Phase 2	60

Table 2 Representative MSC-based trials for digestive disease treatment—cont'd

Disease	NCT number	Status	Phases	Enrollment
Crohn's Disease	NCT04939337	Enrolling	Phase 1	24
IBD	NCT05003947	Recruiting	Phase 1	15
Crohn's Disease	NCT05039411	Recruiting	Phase 1	7
Fistula	NCT05075811	Recruiting	Phase 1, 2	20
Liver Cirrhosis	NCT05080465	Active	Phase 3	700
Liver Cirrhosis	NCT05106972	Recruiting	Not Applicable	30
Decompensated Cirrhosis	NCT05121870	Recruiting	Phase 2	240
Crohn's Disease, IBD	NCT05130983	Recruiting	Phase 1	10
Alcoholic Cirrhosis	NCT05155657	Recruiting	Phase 1	36
Ulcerative Colitis	NCT05176366	Recruiting	Phase 1	10
Crohn's Disease	NCT05177003	Active		27
Decompensated Cirrhosis	NCT05224960	Not recruiting	Phase 2	240
Decompensated Cirrhosis	NCT05227846	Recruiting	Phase 1	9
Crohn's Disease	NCT05262829	Not recruiting	Not Applicable	40
Crohn's Disease	NCT05322057	Completed		14
Crohn's Disease	NCT05330416	Recruiting	Not Applicable	108
Liver Cirrhosis	NCT05331872	Recruiting	Phase 1	20
Rectal Fistula	NCT05407766	Withdrawn	Phase 1	0
Perianal Fistula	NCT05499156	Active	Phase 1, 2	80
Cirrhosis Due to Hepatitis B	NCT05507762	Recruiting	Phase 1, 2	20
Crohn's Disease	NCT05521672	Recruiting	Phase 1, 2	20
Crohn's Disease, UC, IBD	NCT05578508	Withdrawn	Phase 1	0
IBD	NCT05602116	Not recruiting	Phase 1	10

2.3 The application of MSCs in neurological diseases

At present, the incidence of neurodegenerative diseases is increasing worldwide due to the deficiency of effective remedies [15–17]. To date, MSCs have been indicated in the management of various neurological disorders, including spinal cord injury, stroke, traumatic brain injury, Parkinson's disease (PD), multiple sclerosis (MS), Alzheimer's disease (AD), amyotrophic lateral sclerosis (ALS), and Huntington's disease [18–21]. For example,

Reza-Zaldivar et al. assumed that MSCs could alleviate Parkinson's disease via blocking neuronal apoptosis, increasing neurogenesis and synaptogenesis, and promoting angiogenesis [22]. As to hypertensive intracerebral hemorrhage (HICH), Ding et al. demonstrated the ameliorative effect of MSCs, mainly via alleviating the manifestation of hypertensive cerebral hemorrhage [23]. In 2014, Xu et al. took advantage of the well-established experimental autoimmune encephalitis (EAE) model to assess the feasibility of hESC-MSCs for treating multiple sclerosis, and found that the manifestations and demyelination of the mice were collectively ameliorated [24]. Additionally, investigators in the field have also assessed the feasibility of various neurological diseases' management by utilizing MSCs of different origins in animal models such as craniocerebral trauma, cerebral palsy, and spinocerebellar ataxia (SCA) [25–27] (Table 3).

2.4 The application of MSCs in endocrine diseases

From the perspective of developmental biology, the endocrine system is the cell population developed from the embryonic endoderm and mesoderm, which can cooperate with the ectoderm-derived nervous system to maintain the balance and homeostasis of the internal environment of the body. Dysfunction of the endocrine system often leads to the occurrence and progression of a variety of diseases, such as diabetes and its complications (diabetic foot, diabetic nephropathy, diabetic lower limb ischemia, diabetic eye disease), osteoporosis, hyperthyroidism, obesity, hyperuricemia, and gout. Currently, MSC-based cytotherapy is considered to be an efficient treatment for type I/II diabetes mellitus via modulating the pancreatic β cells [28,29]. Interestingly, Lv and colleagues verified that depletion of the diabetic gut microbiota resistance could enhance MSC-based cytotherapy in type I diabetes mellitus [30]; whereas, Jiang et al. demonstrated the safety and efficacy of placenta-derived mesenchymal stem cells (P-MSCs) upon 10 patients with type II diabetes mellitus [31]. Of note, researchers and clinicians have also highlighted the potential application of MSCs in diabetic retinopathy [32,33], osteoporosis [34,35], and hyperuricemia [36] (Table 4).

Table 3 Representative MSC-based trials for neurological disease treatment.

Disease	NCT number	Status	Phases	Enrollment
Multiple Sclerosis	NCT00395200	Completed	Phase 1, 2	10
Multiple Sclerosis	NCT00781872	Completed	Phase 1, 2	24
Multiple Sclerosis	NCT00813969	Completed	Phase 1	24
Chronic Spinal Cord Injury	NCT00816803	Completed	Phase 1, 2	80
Ischemia, Stroke	NCT00875654	Completed	Phase 2	31
Parkinson's Disease	NCT00976430	Terminated	Not Applicable	5
ALS	NCT01051882	Completed	Phase 1, 2	12
Demyelinating Diseases	NCT01056471	Completed	Phase 1, 2	30
Stroke	NCT01091701	Withdrawn	Phase 1, 2	0
ALS	NCT01142856	Completed	Phase 1	1
Spinal Cord Injury	NCT01162915	Suspended	Phase 1	10
Multiple Sclerosis	NCT01228266	Terminated	Phase 2	9
Spinal Cord Injury	NCT01274975	Completed	Phase 1	8
Ischemic Stroke	NCT01297413	Completed	Phase 1, 2	38
Intracerebral Hemorrhage	NCT01298830	Terminated	Phase 1, 2	11
Spinal Cord Injury	NCT01325103	Completed	Not Applicable	14
Hereditary Ataxia	NCT01360164	Unknown status	Phase 1, 2	20
Progressive Multiple Sclerosis	NCT01364246	Unknown status	Phase 1, 2	20
Multiple Sclerosis	NCT01377870	Completed	Phase 1, 2	22
Spinal Cord Injuries	NCT01393977	Unknown status	Phase 2	60
Parkinson's Disease	NCT01446614	Unknown status	Phase 1, 2	20
Spinal Cord Injury	NCT01446640	Unknown status	Phase 1, 2	20
Middle Cerebral Artery Infarction	NCT01461720	Unknown status	Phase 2	50
Stroke	NCT01468064	Completed	Phase 1, 2	20

Continued

Table 3 Representative MSC-based trials for neurological disease treatment—cont'd

Disease	NCT number	Status	Phases	Enrollment
Hereditary Cerebellar Ataxia	NCT01489267	Unknown status	Phase 2	20
ALS	NCT01494480	Unknown status	Phase 2	30
Alzheimer's Disease	NCT01547689	Unknown status	Phase 1, 2	30
Multiple Sclerosis	NCT01606215	Completed	Phase 1, 2	21
ALS	NCT01609283	Completed	Phase 1	27
Spinal Cord Injury	NCT01624779	Completed	Phase 1	15
Severe Brain Injury	NCT01649700	Completed	Phase 1, 2	1
Spinal Cord Injury	NCT01676441	Terminated	Phase 2, 3	20
Ischemic Stroke	NCT01678534	Completed	Phase 2	19
Spinal Cord Injury	NCT01694927	Unknown status	Phase 2	30
Alzheimer's Disease	NCT01696591	Unknown status		14
Stroke	NCT01714167	Unknown status	Phase 1	30
Stroke, Ischemic	NCT01716481	Unknown status	Phase 3	60
Multiple Sclerosis	NCT01730547	Unknown status	Phase 1, 2	15
Multiple Sclerosis	NCT01745783	Completed	Phase 1, 2	26
ALS	NCT01759784	Withdrawn	Phase 1	0
ALS	NCT01759797	Completed	Phase 1	6
Spinal Cord Injury	NCT01769872	Completed	Phase 1, 2	15
ALS	NCT01771640	Completed	Phase 1	8
ALS	NCT01777646	Completed	Phase 2	14
Ischemic Stroke	NCT01849887	Withdrawn	Phase 1, 2	0
Multiple Sclerosis	NCT01854957	Unknown status	Phase 1, 2	20
Spinal Cord Injury	NCT01873547	Completed	Phase 3	300
Multiple Sclerosis	NCT01895439	Completed	Phase 1, 2	13
Spinal Cord Injury	NCT01909154	Completed	Phase 1	12

Ischemic Stroke	NCT01922908	Withdrawn	Phase 1, 2	0
Cerebral Palsy	NCT01929434	Completed	Phase 3	300
Multiple Sclerosis	NCT01933802	Completed	Phase 1	20
ALS	NCT02017912	Completed	Phase 2	48
Multiple Sclerosis	NCT02034188	Completed	Phase 1, 2	20
Relapsing Multiple Sclerosis	NCT02035514	Completed	Phase 1, 2	9
Alzheimer's Disease	NCT02054208	Completed	Phase 1, 2	46
Amyotrophic Lateral Sclerosis	NCT02116634	Withdrawn	Phase 1, 2	0
Spinal Cord Injury	NCT02152657	Completed	Not Applicable	5
Spinal Cord Injury	NCT02165904	Completed	Phase 1	10
Multiple Sclerosis	NCT02166021	Completed	Phase 2	48
Anoxic Brain Injury	NCT02210624	Unknown status	Not Applicable	4
Spinal Cord Injury	NCT02237547	Withdrawn	Phase 1, 2	0
Multiple Sclerosis	NCT02239393	Completed	Phase 2	31
ALS	NCT02290886	Completed	Phase 1, 2	52
Multiple Sclerosis	NCT02326935	Terminated	Phase 1	2
Spinal Cord Injury	NCT02352077	Unknown status	Phase 1	30
Cerebral Infarction	NCT02378974	Completed	Phase 1, 2	19
Multiple Sclerosis	NCT02403947	Terminated	Phase 1, 2	1
Multiple Sclerosis	NCT02418325	Terminated	Phase 1, 2	69
Spinal Cord Injuries	NCT02481440	Completed	Phase 1, 2	102
Spinal Cord Injury	NCT02482194	Completed	Phase 1	9
ALS	NCT02492516	Completed	Phase 1	19
Progressive Multiple Sclerosis	NCT02495766	Completed	Phase 1, 2	8
Traumatic Brain Injury	NCT02525432	Enrolling by invitation	Phase 2	55
Traumatic Brain Injury	NCT02525432	Enrolling	Phase 2	55
Stroke	NCT02564328	Unknown status	Phase 1	40

Continued

Table 3 Representative MSC-based trials for neurological disease treatment—cont'd

Disease	NCT number	Status	Phases	Enrollment
Spinal Cord Injury	NCT02570932	Completed	Phase 2	10
Spinal Cord Injury	NCT02574572	Unknown status	Phase 1	10
Spinal Cord Injury	NCT02574585	Unknown status	Phase 2	40
Stroke	NCT02580019	Unknown status	Phase 2	2
Multiple Sclerosis	NCT02587715	Unknown status	Phase 1, 2	69
Alzheimer's Disease	NCT02600130	Completed	Phase 1	33
Parkinson's Disease	NCT02611167	Completed	Phase 1	20
Alzheimer's Disease	NCT02672306	Unknown status	Phase 1, 2	16
Spinal Cord Injury	NCT02688049	Unknown status	Phase 1, 2	30
Huntington Disease	NCT02728115	Active	Phase 1	6
Brain Death	NCT02742857	Completed	Phase 1	20
Mandibular Fractures	NCT02755922	Completed	Phase 3	20
Brain Injury	NCT02767817	Unknown status	Phase 1	30
Alzheimer Dementia	NCT02833792	Recruiting	Phase 2	40
Stroke	NCT02849613	Withdrawn	Phase 2, 3	0
Injury of Facial Nerve	NCT02853942	Unknown status	Early Phase 1	100
Spinal Muscular Atrophy	NCT02855112	Unknown status	Phase 1, 2	10
ALS	NCT02881476	Unknown status	Phase 1	30
ALS	NCT02881489	Unknown status	Phase 1	30
Alzheimer's Disease	NCT02899091	Unknown status	Phase 1, 2	24
Acute Traumatic Spinal Cord Injury	NCT02917291	Recruiting	Phase 1, 2	48
ALS	NCT02917681	Unknown status	Phase 1, 2	28
Spinal Cord Injuries	NCT02981576	Completed	Phase 1, 2	14
Motor Neuron Disease	NCT02987413	Completed	Phase 1	3

Spinal Cord Injury	NCT03003364	Completed	Phase 1, 2	10
Multiple Sclerosis	NCT03069170	Unknown status	Phase 1	50
Cerebral Palsy	NCT03078621	Unknown status	Phase 1, 2	50
Alzheimer's Disease	NCT03117738	Completed	Phase 1, 2	21
Alzheimer's Disease	NCT03172117	Unknown status	Phase 1, 2	45
Cerebral Infarction	NCT03176498	Suspended	Phase 1, 2	40
Cerebral Infarction	NCT03186456	Suspended	Phase 1	40
Ocular Corneal Burn	NCT03237442	Unknown status	Phase 1, 2	100
Huntington's Disease	NCT03252535	Completed	Phase 2	35
ALS	NCT03268603	Active	Phase 2	69
ALS	NCT03280056	Completed	Phase 3	263
Corneal Scars and Opacities	NCT03295292	Unknown status	Phase 1	15
ALS	NCT03296501	Unknown status	Phase 1	30
Spinal Cord Injuries	NCT03308565	Completed	Phase 1	10
Multiple Sclerosis	NCT03326505	Completed	Phase 1, 2	60
Multiple Sclerosis	NCT03355365	Active	Phase 2	50
Perinatal Arterial Ischemic Stroke	NCT03356821	Completed	Phase 1, 2	10
Hemorrhagic Stroke	NCT03371329	Completed	Phase 1	9
Spinocerebellar Ataxia	NCT03378414	Not recruiting	Phase 2	45
Cerebrovascular Disorders	NCT03384433	Recruiting	Phase 1, 2	5
Cerebral Palsy	NCT03414697	Unknown status	Not Applicable	44
Cerebral Palsy	NCT03473301	Completed	Phase 1, 2	91
Spinal Cord Injuries	NCT03505034	Unknown status	Phase 2	43
Spinal Cord Injuries	NCT03521323	Unknown status	Phase 2	66
Spinal Cord Injury	NCT03521336	Unknown status	Phase 2	84
Parkinson's Disease	NCT03550183	Enrolling	Phase 1	20

Continued

Table 3 Representative MSC-based trials for neurological disease treatment—cont'd

Disease	NCT number	Status	Phases	Enrollment
Parkinson's Disease	NCT03684122	Active	Phase 1, 2	10
Multiple Sclerosis	NCT03696485	Not recruiting	Phase 1, 2	12
Multiple Sclerosis	NCT03778333	Completed	Phase 1	7
Cerebral Palsy	NCT03795974	Unknown status	Phase 2	108
Multiple Sclerosis	NCT03799718	Completed	Phase 2	23
Multiple Sclerosis	NCT03822858	Not available		
ALS	NCT03828123	Completed	Phase 1, 2	26
Spinal Cord Injury	NCT03925649	No longer available		
Cerebral Palsy	NCT03979898	Completed	Early Phase 1	1
Cerebral Palsy	NCT04029896	No longer available		
Alzheimer's Disease	NCT04040348	Active	Phase 1	6
Traumatic Brain Injury	NCT04063215	Active	Phase 1, 2	24
Spinal Cord Injuries	NCT04064957	No longer available		
Cerebral Infarction	NCT04093336	Recruiting	Phase 1, 2	120
Acute Ischemic Stroke	NCT04097652	Recruiting	Phase 1	9
Spinal Cord Injuries	NCT04213131	Unknown status	Not Applicable	42
Huntington's Disease	NCT04219241	Active	Phase 2, 3	35
Alzheimer's Disease	NCT04228666	Withdrawn	Phase 1, 2	0
Ischemic Stroke	NCT04280003	Recruiting	Phase 2	30
Spinal Cord Injuries	NCT04288934	Completed	Phase 1	20
Cerebral Palsy	NCT04314687	Recruiting	Phase 1, 2	78
Alzheimer's Disease	NCT04388982	Unknown status	Phase 1, 2	9
Acute Stroke	NCT04434768	Recruiting	Phase 1	14
Alzheimer's Disease	NCT04482413	Not recruiting	Phase 2	100
Parkinson's Disease	NCT04506073	Active	Phase 2	45
ALS	NCT04514952	No longer available		

Spinal Cord Injuries	NCT04520373	Recruiting	Phase 2	40
Ischemic Stroke	NCT04590118	Recruiting	Phase 1, 2	60
ALS	NCT04651855	Active	Phase 1, 2	20
ALS	NCT04681118	No longer available		
Neurodegenerative Disorders	NCT04684602	Recruiting	Phase 1, 2	5000
Postconcussion Syndrome	NCT04744051	Active	Phase 1	20
ALS	NCT04745299	Recruiting	Phase 3	115
Multiple Sclerosis	NCT04749667	Recruiting	Phase 1, 2	18
Parkinson's Disease	NCT04772378	Available		
Ischemic Stroke	NCT04811651	Recruiting	Phase 2	200
ALS	NCT04821479	Completed	Phase 1, 2	20
Multiple Sclerosis	NCT04823000	Completed	Phase 1, 2	24
Primary Lateral Sclerosis	NCT04825613	Not available		
Alzheimer's Disease	NCT04855955	No longer available		
Multiple System Atrophy	NCT04876326	Recruiting	Not Applicable	15
Stroke	NCT04907188	No longer available		
Parkinson's Disease	NCT04928287	Active	Phase 2	24
Stroke	NCT04953663	Recruiting	Phase 1, 2	60
Alzheimer's Disease	NCT04954534	Not recruiting	Not Applicable	9
Multiple Sclerosis	NCT04956744	Recruiting	Phase 1	30
Parkinson's Disease	NCT04995081	Recruiting	Phase 2	60
Multiple Sclerosis	NCT05003388	Recruiting	Phase 1	15
ALS	NCT05003921	Suspended	Phase 1	20
Ischemic Stroke	NCT05008588	Recruiting	Phase 1, 2	15
Spinal Cord Injuries	NCT05018793	Suspended	Phase 1	15
Cerebral Palsy	NCT05018819	Recruiting	Phase 1	15

Continued

Table 3 Representative MSC-based trials for neurological disease treatment—cont'd

Disease	NCT number	Status	Phases	Enrollment
Traumatic Brain Injury	NCT05018832	Not recruiting	Phase 1	20
Cerebral Palsy	NCT05018871	Suspended	Phase 1	15
Idiopathic Parkinson's Disease	NCT05094011	Not recruiting	Phase 1	9
Multiple Sclerosis	NCT05116540	Recruiting	Phase 2	24
Traumatic Optic Neuropathy	NCT05147701	Recruiting	Phase 1	20
Spinal Cord Injuries	NCT05152290	Recruiting	Phase 1	20
Parkinson's Disease	NCT05152394	Not recruiting	Phase 1	20
Stroke	NCT05158101	Recruiting	Phase 1	15
Ischemic Stroke	NCT05292625	Recruiting	Phase 1, 2	48
Acute Ischemic Stroke	NCT05522569	Not available		
Multiple Sclerosis	NCT05532943	Not recruiting	Phase 1, 2	41

Table 4 Representative MSC-based trials for endocrine disease treatment.

Disease	NCT number	Status	Phases	Enrollment
T1DM	NCT00646724	Unknown status	Phase 1, 2	30
T1DM	NCT00690066	Completed	Phase 2	63
Diabetic Wounds, Venous Stasis Wounds	NCT00815217	Unknown status	Not Applicable	250
Diabetic Foot	NCT00955669	Completed	Phase 1	40
Diabetic Foot	NCT01065337	Completed	Phase 2	30
T1DM	NCT01068951	Completed	Not Applicable	20
T2DM	NCT01142050	Unknown status	Phase 1	24
T1DM	NCT01143168	Unknown status	Phase 1	24
Osteoporosis, Osteopenia	NCT01152580	Completed	Phase 1	19
Diabetic Foot, Critical Limb Ischemia	NCT01216865	Unknown status	Phase 1, 2	50
T1DM	NCT01219465	Unknown status	Phase 1, 2	50
Critical Limb Ischemia	NCT01257776	Completed	Phase 1, 2	33
DM	NCT01322789	Unknown status	Phase 1, 2	10
T1DM	NCT01374854	Unknown status	Phase 1, 2	44
T2DM	NCT01413035	Unknown status	Phase 1, 2	30
T1DM	NCT01496339	Unknown status	Phase 1, 2	50
Osteoporotic Fractures	NCT01532076	Terminated	Phase 2	8
TT1DM, T2DM	NCT01686139	Unknown status	Phase 1	12
T2DM	NCT01719640	Completed	Phase 1, 2	22
T2DM	NCT01759823	Completed	Phase 2, 3	30
T2DM, Osteoporosis	NCT01846533	Unknown status		1200
T2DM	NCT01954147	Unknown status	Phase 1, 2	100
T1DM	NCT01967186	Unknown status	Not Applicable	36
T1DM	NCT02057211	Terminated	Phase 2	10

Continued

Table 4 Representative MSC-based trials for endocrine disease treatment—cont'd

Disease	NCT number	Status	Phases	Enrollment
T1DM	NCT02138331	Unknown status	Phase 2, 3	20
T2DM	NCT02286128	Completed		75
Peripheral Arterial Disease	NCT02287831	Unknown status	Phase 1, 2	30
Critical Limb Ischemia	NCT02287974	Terminated	Phase 1, 2	20
T2DM	NCT02302599	Completed	Phase 2	103
Diabetic Foot, Lower Limb Ischemia	NCT02304588	Unknown status	Phase 1	20
Chronic Pancreatitis	NCT02384018	Completed	Phase 1	3
Diabetic Peripheral Neuropathy	NCT02387749	Completed	Not Applicable	10
Diabetic Foot Ulcer	NCT02394886	Completed	Phase 1	5
Osteoporosis, Spinal Fractures	NCT02566655	Completed	Phase 1	10
T1DM, T2DM	NCT02579148	Withdrawn	Phase 1	0
Diabetic Foot Ulcer	NCT02619877	Completed	Phase 2	59
Diabetic Foot Ulcers	NCT02672280	Unknown status	Phase 1, 2	30
T1DM, T2DM	NCT02745808	Unknown status	Phase 1	30
Diabetic Ketoacidosis	NCT02763423	Unknown status	Phase 2	30
Peripheral Vascular Disease, Diabetic Foot	NCT02796079	Unknown status	Phase 1	240
Peripheral Vascular Disease	NCT02831075	Unknown status	Phase 1	240
Peripheral Vascular Disease, Diabetic Foot	NCT02834858	Unknown status	Phase 1	240
T2DM	NCT02886884	Completed	Phase 1, 2	16
T1DM	NCT02893306	Unknown status	Phase 2	10
Gestational Diabetes Mellitus	NCT02926079	Completed		18
T1DM	NCT02940418	Unknown status	Phase 1	20
T2DM	NCT02945449	Completed	Phase 1	9
Diabetic Foot Ulcer	NCT03183726	Completed		4
Diabetic Foot Ulcer	NCT03183804	Unknown status		54
Osteopenia, Osteoporosis, Osteonecrosis	NCT03197623	Completed	Phase 1	58

Diabetic Foot Ulcer	NCT03248466	Unknown status	Early Phase 1	60
Diabetic Neuropathic Ulcer	NCT03267784	Completed	Phase 1, 2	23
Diabetic Foot	NCT03276312	Completed	Not Applicable	112
Diabetic Nephropathies	NCT03288571	Unknown status	Phase 1, 2	20
CKD, Diabetic Nephropathies	NCT03325322	Enrolling	Phase 2	30
T2DM	NCT03343782	Completed	Phase 1, 2	30
T1DM	NCT03361631	Unknown status	Phase 1	13
Diabetic Foot Ulcer	NCT03370874	Unknown status	Phase 3	164
T2DM	NCT03470415	Unknown status		100
T1DM	NCT03484741	Unknown status	Phase 1, 2	15
T2DM	NCT03751735	Completed	Phase 1, 2	9
Diabetic Foot Ulcer	NCT03754465	Recruiting	Phase 2	56
Diabetic Kidney Disease, Diabetic Nephropathies	NCT03840343	Recruiting	Phase 1	30
Diabetic Foot Ulcer	NCT03865394	Completed	Phase 1, 2	46
T1DM	NCT03912480	Unknown status	Early Phase 1	24
T1DM	NCT03920397	Completed	Not Applicable	30
T2DM	NCT03943940	Unknown status	Phase 1, 2	60
T1DM	NCT03973827	Active	Phase 1, 2	15
T1DM	NCT04061746	Recruiting	Phase 1	60
T1DM	NCT04078308	Unknown status	Phase 1, 2	20
Diabetic Foot Ulcer	NCT04104451	Active	Phase 1	16
Diabetic Nephropathy	NCT04125329	Recruiting	Early Phase 1	15
T2DM	NCT04126603	Recruiting	Phase 3	40
T2DM	NCT04216849	Unknown status	Phase 1, 2	54
T2DM	NCT04441658	Unknown status	Phase 1, 2	30
Diabetic Foot Ulcer	NCT04464213	Recruiting	Phase 1	43
Limb Ischemia, Diabetic Foot	NCT04466007	Recruiting	Phase 2	90

Continued

Table 4 Representative MSC-based trials for endocrine disease treatment—cont'd

Disease	NCT number	Status	Phases	Enrollment
Diabetic Foot Ulcer	NCT04497805	Recruiting	Phase 2	64
T2DM	NCT04501341	Unknown status	Phase 1, 2	15
Osteoporosis	NCT04501354	Unknown status	Phase 2	5
Diabetic Nephropathy	NCT04562025	Unknown status	Not Applicable	38
Diabetic Foot Ulcer	NCT04569409	Unknown status	Phase 3	104
T2DM	NCT04642911	Active		91
Diabetes Complications	NCT04684602	Recruiting	Phase 1, 2	5000
Critical Limb Ischemia	NCT04746599	Recruiting	Not Applicable	20
Diabetes Mellitus, Ischemic Heart Disease	NCT04776239	Recruiting	Phase 1, 2	30
Osteoporosis	NCT04864509	Not recruiting	Phase 4	40
CKD, T1DM, T2DM, Diabetic Nephropathies	NCT04869761	Recruiting	Phase 1	40
Erectile Dysfunction with DM	NCT04972890	Recruiting	Phase 2, 3	12
CKD, Obesity	NCT04998461	Not recruiting		60
Diabetes	NCT05003908	Recruiting	Phase 1	20
Osteoarthritis of Knee	NCT05081921	Recruiting	Phase 1, 2	200
Diabetic Retinopathy	NCT05147701	Recruiting	Phase 1	20
Osteoporosis	NCT05152381	Suspended	Phase 1	20
Diabetic Foot Ulcer, Cutaneous Ulcer	NCT05165628	Recruiting	Phase 1	30
Diabetes Mellitus	NCT05207995	Not recruiting	Phase 1, 2	30
Diabetic Foot	NCT05243368	Not recruiting	Not Applicable	30
Diabetic Foot Ulcer	NCT05274295	Recruiting	Not Applicable	5
T1DM	NCT05308836	Recruiting	Phase 1	10
Diabetic Peripheral Neuropathy	NCT05507697	Recruiting	Phase 1, 2	42
Diabetic Foot Ulcer	NCT05595681	Not recruiting	Phase 1	30
Chronic Limb Ischemia	NCT05631444	Completed	Phase 1, 2	24

2.5 The application of MSCs in immune diseases

As mentioned above, MSCs possess unique immunomodulatory properties, which provides them with broad prospects in both preclinical and clinical practice for immune disease relief, such as systemic sclerosis (SSc), systemic lupus erythematosus (SLE), graft-versus-host disease (aGvHD), and allergic diseases (e.g., pediatric asthma, allergic rhinitis) [37–40]. For example, Christopeit et al., for the first time, reported the remission of an SSc case with refractory severe syndrome via infusion of allogeneic BM-MSCs [41]. In 2015, Thiel et al. demonstrated the ameliorative effect of hESC-MSCs on lupus nephritis mice, which could efficiently delay disease progression, protect kidney function, and prolong the survival of mice [42]. Wen et al. reported the alleviative effect of allogeneic MSCs for 69 SLE cases, and confirmed the attenuated SLE Disease Activity Index (SLEDAI) [43]. Of note, Peltzer and colleagues summarized the progress of autologous or allogeneic MSCs for treatment of a variety of autoimmune diseases in clinical practice, including SSc, CD, and SLE [39,44] (Table 5).

2.6 The application of MSCs in movement diseases

Osteoarthritis (OA) is a typical representative of motor system diseases, and is associated with inflammation, infection, trauma, autoimmune reaction, metabolic disorders, degenerative diseases, and other factors [45,46]. Due to the limitations in efficacy and relapse after receiving traditional remedies, pioneering investigators in the field have turned to MSC-based novel cytotherapy. For example, Jo et al. conducted a clinical project on 18 patients with knee osteoarthritis, and verified the safety and efficacy of autologous adipose tissue-derived MSCs (AD-MSCs), including reduced cartilage defects of the medial femoral condyle and medial tibial condyle, relieved pain, and hyaline cartilage regeneration [47]. Recently, we assessed the therapeutic effect of various MSCs/biomaterial composites (hyaluronic acid, alginate) on rabbits with knee osteoarthritis [48]. In 2010, Lian et al. verified the preferable outcome of mice with posterior limb ischemia treated with iPSC-MSCs rather than BM-MSCs for blood vessel and muscle regeneration [49]. Notably, substantive studies have also indicated the amative effect of bone grafting pith decompression and UC-MSCs combination for femoral head necrosis administration [50] (Table 6).

Table 5 Representative MSC-based trials for immune disease treatment.

Disease	NCT number	Status	Phases	Enrollment
Systemic Sclerosis	NCT05016804	Recruiting	Phase 1	20
Systemic Sclerosis	NCT00962923	Unknown status	Phase 1, 2	20
Systemic Scleroderma	NCT02213705	Completed	Phase 1, 2	20
Systemic Sclerosis	NCT02975960	Completed	Not Applicable	7
Systemic Sclerosis	NCT04356287	Not recruiting	Phase 1, 2	18
Systemic Sclerosis	NCT03060551	Completed	Early Phase 1	20
Systemic Sclerosis	NCT03211793	Recruiting	Phase 1, 2	20
Systemic Scleroderma	NCT03629002	Unknown status		30
Systemic Sclerosis, Dermatomyositis	NCT03582800	Recruiting	Phase 2	40
Critical Limb Ischemia, Peripheral Arterial Disease	NCT04746599	Recruiting	Not Applicable	20
Multiple Sclerosis	NCT01745783	Completed	Phase 1, 2	26
SLE	NCT04184258	Completed	Phase 1, 2	7
SLE	NCT03219801	Unknown status	Early Phase 1	10
SLE	NCT02633163	Recruiting	Phase 2	81
Refractory SLE	NCT00698191	Unknown status	Phase 1, 2	20
SLE	NCT01741857	Unknown status	Phase 1, 2	40
SLE	NCT03171194	Completed	Phase 1	6
Lupus Erythematosus	NCT03562065	Recruiting	Phase 1, 2	10
Lupus Nephritis	NCT05631717	Recruiting	Phase 3	40
SLE	NCT04835883	Recruiting	Phase 2	10
Lupus Nephritis	NCT03174587	Completed	Phase 1	7
Lupus Nephritis	NCT03673748	Not recruiting	Phase 2	20
Lupus Nephritis	NCT04522505	Active		7
Allergic Rhinitis	NCT05167552	Not recruiting	Phase 1, 2	60

Table 5 Representative MSC-based trials for immune disease treatment—cont'd

Disease	NCT number	Status	Phases	Enrollment
Allergic Rhinitis	NCT05151133	Recruiting	Phase 1	18
Lupus Nephritis	NCT03580291	Unknown status	Phase 2	230
Lupus Nephritis	NCT00659217	Unknown status	Phase 1, 2	20
Lupus Nephritis	NCT04318600	Completed	Phase 1	16
Lupus Nephritis, Lupus Erythematosus	NCT03673748	Not recruiting	Phase 2	20
Lupus Nephritis	NCT03458156	Unknown status	Not Applicable	30
Lupus Nephritis	NCT03174587	Completed	Phase 1	7
Lupus Nephritis	NCT05631717	Recruiting	Phase 3	40
Lupus Nephritis	NCT01539902	Unknown status	Phase 2	25
Lupus Nephritis	NCT04522505	Active		7
SLE	NCT04835883	Recruiting	Phase 2	10
Refractory SLE	NCT00698191	Unknown status	Phase 1, 2	20

Table 6 Representative MSC-based trials for movement disease treatment.

Disease	NCT number	Status	Phases	Enrollment
Knee Osteoarthritis	NCT05160831	Not recruiting	Not Applicable	50
Knee Osteoarthritis	NCT03956719	Unknown status	Not Applicable	8
Osteoarthritis	NCT01504464	Completed	Phase 2	40
Knee Osteoarthritis	NCT03383081	Recruiting	Phase 2	60
Knee Osteoarthritis	NCT04351932	Unknown status	Phase 3	54
Osteoarthritis	NCT01459640	Unknown status	Phase 2	50
Osteoarthritis	NCT03164083	Withdrawn	Phase 2	0
Knee Osteoarthritis	NCT04130100	Unknown status	Early Phase 1	60
Knee Osteoarthritis	NCT03800810	Unknown status	Early Phase 1	9
Knee Osteoarthritis	NCT01809769	Completed	Phase 1, 2	18
Osteoarthritis	NCT01207661	Completed	Phase 1	6
Hip Osteoarthritis	NCT01499056	Completed	Phase 1	6
Knee Osteoarthritis	NCT02544802	Completed	Phase 1	4
Knee Osteoarthritis	NCT03357575	Unknown status	Not Applicable	14
Knee Osteoarthritis	NCT02237846	Withdrawn	Phase 1, 2	0
Knee Osteoarthritis	NCT03166865	Unknown status	Phase 1, 2	60
Knee Osteoarthritis	NCT04208646	Unknown status	Phase 2	108
Knee Osteoarthrosis	NCT02963727	Recruiting	Phase 1	10
Knee Osteoarthritis, Osteoarthritis	NCT03869229	Unknown status	Phase 1, 2	100
Osteoarthritis	NCT02641860	Completed	Phase 1	22
Knee Osteoarthritis	NCT03969680	Recruiting	Not Applicable	60
Osteoarthritis, Arthritis of Knee, Knee Osteoarthritis	NCT01586312	Completed	Phase 1, 2	30
Knee Osteoarthritis	NCT01183728	Completed	Phase 1, 2	12
Knee Osteoarthrosis	NCT02966951	Recruiting	Phase 1	10
Knee Osteoarthritis	NCT01985633	Unknown status	Phase 1, 2	24

Knee Osteoarthritis	NCT04212728	Recruiting	Not Applicable	60
Knee Osteoarthritis	NCT01453738	Completed	Phase 2	60
Osteoarthritis	NCT02123368	Completed	Phase 1, 2	30
Knee Osteoarthritis	NCT03602872	Withdrawn	Phase 1	0
Knee Osteoarthrosis	NCT04314661	Recruiting	Phase 1, 2	20
Knee Osteoarthrosis	NCT04326985	Completed	Early Phase 1	20
Knee Osteoarthritis	NCT02365142	Unknown status	Phase 1, 2	38
Knee Osteoarthritis	NCT01448434	Completed	Phase 2	72
Knee Osteoarthritis	NCT03358654	Unknown status	Not Applicable	9
Knee Osteoarthritis	NCT05579665	Active	Phase 1, 2	45
Osteoarthritis	NCT01895413	Completed	Phase 1, 2	10
Knee Osteoarthritis	NCT02162693	Completed	Phase 2	53
Osteoarthritis	NCT01436058	Completed	Phase 1	6
Hip/Knee/Glenohumeral Osteoarthritis	NCT03866330	Unknown status	Phase 1, 2	100
Knee Osteoarthritis, Cartilage Injury	NCT03477942	Recruiting	Phase 1	16
Knee Osteoarthritis	NCT02003131	Withdrawn	Phase 1, 2	0
Knee Osteoarthritis	NCT04368806	Recruiting	Phase 2, 3	140
Shoulder/Knee Osteoarthritis	NCT04448106	Not recruiting	Phase 2	300
Knee Osteoarthritis	NCT02958267	Completed	Phase 2	32
Knee Osteoarthritis	NCT05288725	Not recruiting	Phase 1, 2	120
Knee Osteoarthritis	NCT04893174	Recruiting	Phase 1	6
Knee Osteoarthritis	NCT04863183	Not recruiting	Phase 1, 2	30
Knee Osteoarthritis	NCT04520945	Unknown status	Phase 2	100
Degenerative Arthritis	NCT01300598	Completed	Phase 1, 2	18
Knee Osteoarthritis	NCT00850187	Completed	Phase 1	6
Knee Osteoarthritis	NCT03589287	Completed	Phase 1, 2	18

Continued

Table 6 Representative MSC-based trials for movement disease treatment—cont'd

Disease	NCT number	Status	Phases	Enrollment
Cartilage Damage, Degenerative Osteoarthritis	NCT02776943	Unknown status	Phase 1, 2	20
Knee Osteoarthritis	NCT05016011	Recruiting	Phase 2	50
Knee Osteoarthritis	NCT05349565	Recruiting	Not Applicable	26
Knee Osteoarthritis	NCT03357770	Unknown status	Not Applicable	9
Osteoarthritis, Spinal Arthritis	NCT05147675	Recruiting	Phase 1	20
Knee Osteoarthritis	NCT05086939	Recruiting	Phase 3	120
Knee Osteoarthritis	NCT04240873	Unknown status	Phase 1, 2	24
Rheumatoid Arthritis	NCT01873625	Completed	Phase 2, 3	60
Knee Osteoarthritis	NCT05027581	Recruiting	Phase 2	70
Osteoarthritis	NCT03818737	Completed	Phase 3	480
Cartilage Diseases, Osteoarthritis	NCT02291926	Completed	Phase 1	20
Knee Osteoarthritis	NCT02118519	Completed	Phase 2	13
Knee Osteoarthritis	NCT03955497	Unknown status	Phase 1, 2	30
Knee Osteoarthritis	NCT01879046	Completed	Not Applicable	35
Degenerative Arthritis, Knee Osteoarthritis	NCT03990805	Completed	Phase 3	260
Degenerative Arthritis, Knee Osteoarthritis	NCT03000712	Completed	Not Applicable	26
Knee Osteoarthritis	NCT03509025	Completed	Phase 2	11
Osteoarthritis	NCT03014037	Completed	Not Applicable	35
Knee Osteoarthritis, Musculoskeletal Disease	NCT03337243	Completed	Not Applicable	60
Knee Osteoarthritis	NCT02855073	Unknown status	Phase 2	28
Knee Osteoarthritis	NCT04037345	Completed	Phase 1	12
Knee Osteoarthritis	NCT05344157	Recruiting	Phase 1, 2	54
Knee Osteoarthritis	NCT05182034	Recruiting	Phase 2	90
Knee Osteoarthritis	NCT02674399	Completed	Phase 2	28
Degenerative Arthritis, Osteochondral Defects	NCT00891501	Unknown status	Phase 2, 3	25

Knee Osteoarthritis	NCT03943576	Unknown status	Phase 1, 2	30
Osteoarthritis	NCT03028428	Unknown status	Phase 2	1
Knee Osteoarthritis	NCT04339504	Recruiting	Phase 1	12
Cartilage Degeneration, Knee Osteoarthritis	NCT03648463	Withdrawn	Not Applicable	0
Knee Osteoarthritis	NCT01159899	Unknown status	Early Phase 1	50
Knee Osteoarthritis	NCT04427930	Active	Phase 3	129
Degenerative Arthritis, Knee Arthritis	NCT04825730	Not recruiting	Not Applicable	14
Knee Osteoarthritis	NCT02468492	Completed	Early Phase 1	18
Knee Osteoarthritis	NCT05484856	Completed	Phase 1, 2	140
Knee Osteoarthritis	NCT05280002	Completed	Phase 2	30
Tibia or Femur Pseudo-arthrosis	NCT00557635	Suspended	Phase 2	50
Primary Osteoarthritis of Knee	NCT01931007	Completed	Phase 1	25
Cartilage Injury, Osteoarthritis	NCT01041001	Completed	Phase 3	104
Knee Osteoarthritis	NCT03308006	Unknown status	Phase 2	18
Degenerative Arthritis, Knee Osteoarthritis	NCT02658344	Completed	Phase 2	24
Knee Osteoarthritis	NCT03379168	Active	Not Applicable	75
Knee Pain Chronic, Knee Osteoarthritis	NCT04604288	No longer available		
Cartilage Injury, Knee Osteoarthritis	NCT02696876	Recruiting	Not Applicable	20
Knee Osteoarthritis	NCT04230902	Unknown status	Phase 3	48
Knee Osteoarthritis	NCT05000593	Completed	Not Applicable	30
Osteoarthritis	NCT02580695	Completed	Phase 1, 2	30
Knee Osteoarthritis	NCT03790189	Unknown status	Not Applicable	25
Rheumatoid Arthritis, Knee Osteoarthritis, Knee	NCT03067870	Unknown status	Phase 1	100
Degenerative Osteoarthritis	NCT01626677	Completed	Phase 3	103
Degenerative Arthritis, Knee Arthritis	NCT04821102	Not recruiting	Not Applicable	21
Knee Osteoarthritis	NCT04716803	Recruiting	Not Applicable	10
Osteoarthritis	NCT01926327	Completed	Phase 3	150

Continued

Table 6 Representative MSC-based trials for movement disease treatment—cont'd

Disease	NCT number	Status	Phases	Enrollment
Knee Osteoarthritis	NCT03788265	Unknown status	Not Applicable	60
Knee Osteoarthritis	NCT04234412	Unknown status	Not Applicable	10
Knee Osteoarthritis	NCT02351011	Completed	Phase 1, 2	12
Osteoarthritis Postmeniscectomy	NCT02582489	Recruiting	Not Applicable	100
Hip Osteoarthritis	NCT04990128	Not recruiting	Phase 3	100
Knee Osteoarthritis	NCT03048773	Completed	Not Applicable	40
Osteoarthritis, Knee Injuries	NCT01227694	Completed	Phase 1, 2	15
Knee Osteoarthritis	NCT02964143	Unknown status	Not Applicable	306
Knee Osteoarthritis	NCT05447767	Recruiting	Phase 2	240
Knee Osteoarthritis	NCT04749758	Completed	Not Applicable	66
Knee Osteoarthritis, Cartilage Degeneration	NCT04310852	Unknown status		25
Hip Osteoarthritis	NCT03410355	Terminated	Not Applicable	6
Knee Osteoarthritis	NCT04453111	Unknown status	Phase 1, 2	45
Shoulder Arthritis	NCT04308213	Recruiting	Not Applicable	30
Tendinopathy, Osteoarthritis	NCT05660824	Not recruiting	Phase 4	195
Temporomandibular Disorder	NCT05305833	Recruiting	Phase 1, 2	20
Knee Osteoarthritis	NCT04043819	Unknown status	Phase 1	125
Knee Osteoarthritis	NCT05081921	Recruiting	Phase 1, 2	200
Osteoarthritis	NCT01739504	Terminated	Not Applicable	10
Degenerative Osteoarthritis, Rheumatoid Arthritis	NCT01413061	Completed	Not Applicable	140
Posttraumatic Osteoarthritis of Knee	NCT04222140	Unknown status	Not Applicable	40
Osteoarthritis	NCT04223622	Recruiting		24
Osteoarthritis	NCT01585857	Completed	Phase 1	18
Hip Osteoarthritis	NCT03608579	Recruiting	Phase 1	24
Osteoarthritis	NCT02838069	Active	Phase 2	100

Osteoarthritis	NCT01038596	Completed		30
Degeneration Articular Cartilage Knee	NCT01733186	Completed	Phase 1, 2	12
Avascular Necrosis of the Femoral Head	NCT02065167	Completed	Phase 2	26
Avascular Necrosis of the Femoral Head	NCT01643655	Completed	Not Applicable	15
Avascular Necrosis of the Femoral Head	NCT01605383	Completed	Phase 1, 2	23
Osteonecrosis of the Femoral Head	NCT03180463	Suspended	Phase 1, 2	30
Necrosis	NCT00821470	Completed	Phase 1	21
Osteonecrosis of the Femoral Head	NCT01700920	Completed	Phase 2	3
Avascular Necrosis of the Femoral Head	NCT02890537	Completed	Phase 2	82
Osteochondritis of the Femoral Head	NCT00813267	Unknown status	Early Phase 1	15
Osteonecrosis of the Femoral Head	NCT01529008	Terminated	Phase 3	68
Osteonecrosis	NCT03269409	Terminated	Phase 1	8
Nontraumatic Osteonecrosis of the Femoral Head	NCT01544712	Completed	Not Applicable	50

2.7 The application of MSCs in respiratory diseases

To date, MSCs-based cytotherapy has been involved in management of a number of respiratory diseases, such as pneumonia, acute lung injury (ALI), acute respiratory distress syndrome (ARDS), pulmonary fibrosis, asthma, and even coronavirus disease 2019 (COVID-19) [51–54].

As one of the major etiological factors in acute respiratory failure, ALI reveals pulmonary manifestations (e.g., diffuse alveolar damage), and the mortality rate of ALI patients is 40% [55,56]. State-of-the-art preclinical and clinical studies have suggested the potential of MSCs for ALI treatment. For example, using an influenza virus-induced ALI mouse model and pig model, Bouvier et al. and Khatri et al. verified the ameliorative effect of MSCs and MSC-derived extracellular vesicle (MSC-EV) for ALI management via antiinfluenza and antiinflammatory effects, as well as restraining virus replication and suppressing cell apoptosis [57–59]. Another two studies further demonstrated the efficacious attenuation of a lipopolysaccharide (LPS)-induced ALI sheep or mouse model via intrapulmonary delivery of BM-MSCs or UC-MSCs, respectively [56,60,61]. It is noteworthy that hESC-MSCs seemingly showed minimal improving effect upon pulmonary edema reduction or lung protein permeability compared to BM-MSCs [62].

Distinguish from those with ALI, patients with ARDS present typical clinical manifestations such as severe hypoxemia and diffuse alveolar damage, which has been recognized as the most common cause of respiratory failure, and results in a mortality rate of 40% due to the deficiency of effective treatment regimens [63,64]. For decades, a large number of preclinical studies have demonstrated the involvement of MSC-based therapeutics for ARDS administration based on the infectious ARDS model and sterile inflammatory ARDS model [64]. Meanwhile, Zheng et al. and Matthay et al., respectively, reported the clinical outcomes of ARDS patients with AD-MSCs and BM-MSCs infusion, and confirmed the reliable safety and effectiveness based on multiple clinical parameters [65]. Very recently, we and other investigators have put forward the feasibility of MSC-based treatment regimens and override precedence for COVID-19-associated pneumonia and ARDS [51,52,56,66].

Pulmonary fibrosis is an intractable and debilitating lung disease with multidimensional manifestations such as interstitial fibrosis, hypoxemic respiratory failure, and disruption of pulmonary function [67,68]. In recent years, preclinical and clinical investigations have highlighted the potential of MSCs and its derivations in the management of pulmonary fibrosis, including idiopathic pulmonary fibrosis (IPF) [69]. For instance, Srour and Thébaud systematically summarized the efficacy and molecular mechanism of MSCs on bleomycin-induced pulmonary fibrosis animal models [70]. Interestingly, Chen et al. and Willis et al. verified the ameliorative effect of AD-MSCs and AD-MSC-derived exosomes (AD-MSC-Exo) upon pulmonary fibrosis and experimental bronchopulmonary dysplasia during rat silicosis [71,72].

In recent years, investigators have also indicated the involvement of MSC-based remedies for treatment of other respiratory diseases. In 2012, Fu et al. found that iPSC-MSC transplantation was adequate to prevent a variety of allergic trachea inflammations such as asthma and allergic rhinitis [73]. Subsequently, Abreu et al. took advantage of the ovalbumin-induced C57BL/6 asthma model, and verified that airway hyperresponsiveness, airway obstruction, and asthma symptoms were significantly improved in mice after BM-MSC treatment [74]. Another study, on 20 cases with severe asthma, showed that BM-MSC infusion was adequate to inhibit the proliferation of peripheral T lymphocytes (TLCs), which was consistent with the report by Shin et al. that lung type 2 inflammation could be effectively attenuated by UC-MSC treatment [75]. As to chronic obstructive pulmonary disease (COPD), there were 6 clinical studies on BM-MSCs-based cytotherapy upon moderate–severe COPD patients that, collectively, indicated the feasibility of MSC infusion for eliminating trachea and bronchospasm, enhancing the equilibrium ability of ciliary cells and epithelial cells [40,76] (Table 7).

2.8 The application of MSCs in urogenital diseases

Recent advancements in MSC-based technology have highlighted the feasibility of exploring novel remedies for urogenital disease administration. MSCs of different origins have been recognized as splendid multipotent progenitor cells with unique self-renewal and multilineage differentiation capacity into mesenchymal lineages, and thus provide new paradigms for the treatment of intractable diseases including reproductive disorders, urinary diseases, neurological disorders, and metabolic/endocrine-associated diseases [77]. Nowadays, as a pivotal branch of regenerative medicine, MSC-based cytotherapy has become a prominent treatment modality for regenerative medicine [78].

Reproductive disorders represent a great threat to men's and women's reproductive health as well as contributing to infertility. For instance,

Table 7 Representative MSC-based trials for respiratory disease treatment.

Disease	NCT number	Status	Phases	Enrollment
ARDS	NCT01775774	Completed	Phase 1	9
ALI, ARDS	NCT02444455	Unknown status	Phase 1, 2	20
ARDS	NCT04390152	Recruiting	Phase 1, 2	40
ARDS	NCT02804945	Completed	Phase 1	20
ARDS	NCT03608592	Unknown status	Not Applicable	26
ARDS	NCT04289194	Recruiting	Phase 1, 2	26
ARDS	NCT04602104	Unknown status	Phase 1, 2	169
ALI	NCT04951882	Recruiting	Early Phase 1	30
ARDS	NCT04909879	Withdrawn	Phase 2	0
ARDS	NCT01902082	Unknown status	Phase 1	20
ARDS, COVID-19	NCT05354141	Recruiting	Phase 3	400
COVID-19	NCT04565665	Recruiting	Phase 1, 2	70
COVID-19	NCT04416139	Unknown status	Phase 2	10
ARDS	NCT04456361	Active	Early Phase 1	9
ARDS, COVID-19	NCT05240430	Recruiting		1
ARDS	NCT04347967	Not recruiting	Phase 1	18
COVID-19	NCT04366063	Unknown status	Phase 2, 3	60
ARDS, COVID-19	NCT05387278	Recruiting	Phase 1	20
ARDS, COVID-19	NCT04625738	Completed	Phase 2	30
ARDS	NCT03042143	Recruiting	Phase 1, 2	129
ARDS, COVID-19	NCT04447833	Active	Phase 1	7
ARDS, COVID-19	NCT04355728	Completed	Phase 1, 2	24
ARDS, COVID-19	NCT04798716	Not recruiting	Phase 1, 2	55
ARDS, COVID-19	NCT04452097	Not recruiting	Phase 1, 2	39

ARDS, COVID-19	NCT04490486	Withdrawn	Phase 1	0
ARDS, COVID-19	NCT04390139	Active	Phase 1, 2	30
ARDS	NCT05127122	Not recruiting	Phase 1, 2	81
ARDS, COVID-19	NCT04345601	Recruiting	Phase 1, 2	66
ARDS, COVID-19	NCT04524962	Recruiting	Phase 1, 2	30
ARDS, COVID-19	NCT04537351	Completed	Phase 1, 2	14
ARDS, COVID-19	NCT04377334	Not recruiting	Phase 2	40
ARDS, COVID-19	NCT04371393	Terminated	Phase 3	223
ARDS, COVID-19	NCT04493242	Completed	Phase 2	120
ARDS	NCT04333368	Completed	Phase 1, 2	47
ARDS, COVID-19	NCT04629105	Recruiting	Phase 1	70
ARDS, COVID-19	NCT05092724	Not recruiting	Not Applicable	20
ARDS, COVID-19	NCT04466098	Active	Phase 2	9
ARDS, COVID-19	NCT04400032	Completed	Phase 1, 2	15
ARDS	NCT01775774	Completed	Phase 1	9
ARDS	NCT04390152	Recruiting	Phase 1, 2	40
ARDS	NCT02804945	Completed	Phase 1	20
ARDS	NCT03608592	Unknown status	Not Applicable	26
ARDS	NCT04602104	Unknown status	Phase 1, 2	169
ARDS	NCT04289194	Recruiting	Phase 1, 2	26
ARDS	NCT04909879	Withdrawn	Phase 2	0
ARDS, COVID-19	NCT04565665	Recruiting	Phase 1, 2	70
ARDS	NCT01902082	Unknown status	Phase 1	20
ALI/ARDS	NCT02444455	Unknown status	Phase 1, 2	20
ARDS, COVID-19	NCT05354141	Recruiting	Phase 3	400
COVID-19	NCT04416139	Unknown status	Phase 2	10
ARDS, COVID-19	NCT04456361	Active	Early Phase 1	9

Continued

Table 7 Representative MSC-based trials for respiratory disease treatment—cont'd

Disease	NCT number	Status	Phases	Enrollment
ARDS, COVID-19	NCT05240430	Recruiting		1
ARDS, COVID-19	NCT05387278	Recruiting	Phase 1	20
COVID-19	NCT04366063	Unknown status	Phase 2, 3	60
ARDS	NCT04347967	Not recruiting	Phase 1	18
ARDS, COVID-19	NCT04625738	Completed	Phase 2	30
ARDS	NCT02112500	Unknown status	Phase 2	10
ARDS	NCT05127122	Not recruiting	Phase 1, 2	81
ALI	NCT04951882	Recruiting	Early Phase 1	30
ARDS, COVID-19	NCT04355728	Completed	Phase 1, 2	24
ARDS, COVID-19	NCT04447833	Active	Phase 1	7
ARDS	NCT03042143	Recruiting	Phase 1, 2	129
ARDS, COVID-19	NCT04798716	Not recruiting	Phase 1, 2	55
ARDS, COVID-19	NCT04490486	Withdrawn	Phase 1	0
ARDS, COVID-19	NCT04452097	Not recruiting	Phase 1, 2	39
ARDS	NCT05491681	Not recruiting	Phase 1	9
ARDS, COVID-19	NCT04348461	Suspended	Phase 2	100
COVID-19	NCT04905836	Recruiting	Phase 2	60
ARDS, COVID-19	NCT04728698	Withdrawn	Phase 2	0
ARDS, COVID-19	NCT04390139	Active	Phase 1, 2	30
ARDS, COVID-19	NCT04345601	Recruiting	Phase 1, 2	66
ARDS, COVID-19	NCT04629105	Recruiting	Phase 1	70
ARDS, COVID-19	NCT04524962	Recruiting	Phase 1, 2	30
COVID-19	NCT04903327	Recruiting	Phase 2	100
ARDS, COVID-19	NCT04537351	Completed	Phase 1, 2	14
ARDS, COVID-19	NCT04377334	Not recruiting	Phase 2	40

ARDS, COVID-19	NCT04493242	Completed	Phase 2	120
ARDS, COVID-19	NCT04333368	Completed	Phase 1, 2	47
ARDS, COVID-19	NCT04371393	Terminated	Phase 3	223
ARDS, COVID-19	NCT05092724	Not recruiting	Not Applicable	20
ARDS, COVID-19	NCT04466098	Active	Phase 2	9
ARDS, COVID-19	NCT04400032	Completed	Phase 1, 2	15
IPF	NCT05468502	Recruiting	Phase 1	18
IPF	NCT01919827	Completed	Phase 1	17
Postradiotherapy Pulmonary Fibrosis	NCT02277145	Completed	Phase 1	10
IPF	NCT01385644	Completed	Phase 1	8
IPF	NCT02013700	Terminated	Phase 1	9
IPF	NCT05016817	Recruiting	Phase 1	20
IPF	NCT02135380	Unknown status	Phase 1, 2	60
IPF	NCT02594839	Completed	Phase 1, 2	20
Systemic Sclerosis Pulmonary, Pulmonary Hypertension, Pulmonary Fibrosis	NCT04432545	Available		
COVID-19, Lung Fibrosis	NCT05191381	Recruiting		40
Pulmonary Fibrosis	NCT04541680	Recruiting	Phase 3	250
Asthma	NCT03137199	Terminated	Phase 1	3
Asthma	NCT02192736	Unknown status	Phase 1, 2	20
Pulmonary Disease, Asthma, COPD	NCT05147688	Recruiting	Phase 1	20
COVID-19	NCT04444271	Unknown status	Phase 2	20
COVID-19	NCT05122234	Completed	Phase 3	40
COVID-19	NCT04713878	Completed	Not Applicable	21

Continued

Table 7 Representative MSC-based trials for respiratory disease treatment—cont'd

Disease	NCT number	Status	Phases	Enrollment
COVID-19	NCT04429763	Unknown status	Phase 2	30
COVID-19 Pneumonia	NCT04898088	Completed	Not Applicable	30
COVID-19 Pneumonia	NCT04315987	Completed	Phase 2	90
COVID-19	NCT04416139	Unknown status	Phase 2	10
COVID-19	NCT04313322	Unknown status	Phase 1	5
COVID-19	NCT04252118	Unknown status	Phase 1	20
COVID-19	NCT04611256	Unknown status	Phase 1	20
COVID-19	NCT05501418	Active	Phase 1, 2	75
COVID-19	NCT05125562	Not recruiting	Phase 2	30
COVID-19	NCT04336254	Recruiting	Phase 1, 2	20
COVID-19 Pneumonia	NCT04371601	Active	Early Phase 1	60
COVID-19	NCT04346368	Unknown status	Phase 1, 2	20
ARDS, COVID-19	NCT05240430	Recruiting		1
ARDS, COVID-19	NCT04456361	Active	Early Phase 1	9
COVID-19	NCT04288102	Completed	Phase 2	100
COVID-19	NCT04909892	Withdrawn	Phase 2	0
COVID-19	NCT04349631	Completed	Phase 2	56
COVID-19	NCT04905836	Recruiting	Phase 2	60
ARDS, COVID-19	NCT04625738	Completed	Phase 2	30
COVID-19	NCT04992247	Not recruiting	Phase 2	60
COVID-19	NCT05017298	Not recruiting	Phase 2	30
COVID-19	NCT04753476	Recruiting	Phase 2	48
ARDS, COVID-19	NCT04657458	Available		
ARDS, COVID-19	NCT04728698	Withdrawn	Phase 2	0
COVID-19	NCT04339660	Unknown status	Phase 1, 2	30

COVID-19	NCT04273646	Unknown status	Not Applicable	48
COVID-19	NCT04428801	Not recruiting	Phase 2	200
COVID-19	NCT04366271	Withdrawn	Phase 2	0
COVID-19	NCT05132972	Recruiting	Phase 2, 3	42
COVID-19, Prophylaxis	NCT04573270	Completed	Phase 1	40
COVID-19 Pneumonia	NCT04352803	Not recruiting	Phase 1	20
COVID-19	NCT04302519	Unknown status	Early Phase 1	24
COVID-19	NCT05116761	Not recruiting	Phase 1, 2	60
COVID-19 Pneumonia	NCT04527224	Recruiting	Phase 1, 2	10
ARDS, COVID-19	NCT05354141	Recruiting	Phase 3	400
COVID-19 Pneumonia	NCT04382547	Completed	Phase 1, 2	32
COVID-19 Pneumonia	NCT04457609	Unknown status	Phase 1	40
ARDS, COVID-19	NCT04490486	Withdrawn	Phase 1	0
ARDS, COVID-19	NCT04355728	Completed	Phase 1, 2	24
COVID-19 Pneumonia	NCT04461925	Unknown status	Phase 1, 2	30
COVID-19	NCT04888949	Recruiting	Phase 2	20
ARDS, COVID-19	NCT05387278	Recruiting	Phase 1	20
COVID-19	NCT04366063	Unknown status	Phase 2, 3	60
COVID-19	NCT04362189	Terminated	Phase 2	53
COVID-19	NCT05216562	Recruiting	Phase 2, 3	60
COVID-19	NCT04535856	Completed	Phase 1	9
ARDS, COVID-19	NCT04565665	Recruiting	Phase 1, 2	70
COVID-19	NCT04903327	Recruiting	Phase 2	100
COVID-19	NCT04348435	Completed	Phase 2	53
COVID-19 Pneumonia	NCT05433298	Recruiting	Phase 1, 2	60
COVID-19	NCT04366323	Completed	Phase 1, 2	26
	NCT04493242	Completed	Phase 2	120

Continued

Table 7 Representative MSC-based trials for respiratory disease treatment—cont'd

Disease	NCT number	Status	Phases	Enrollment
ARDS, COVID-19, Pneumonia				
ARDS, COVID-19	NCT04629105	Recruiting	Phase 1	70
ARDS, COVID-19	NCT04494386	Active	Phase 1, 2	17
ARDS, COVID-19	NCT04452097	Not recruiting	Phase 1, 2	39
COVID-19	NCT04780685	Recruiting	Phase 2	40
ARDS, COVID-19	NCT05191381	Recruiting		40
ARDS, COVID-19	NCT04377334	Not recruiting	Phase 2	40
ARDS	NCT04909879	Withdrawn	Phase 2	0
ARDS	NCT04390152	Recruiting	Phase 1, 2	40
COVID-19, Cytokine Release Syndrome, ARDS	NCT04492501	Completed	Not Applicable	600
COVID-19	NCT04293692	Withdrawn	Not Applicable	0
ARDS, COVID-19	NCT04345601	Recruiting	Phase 1, 2	66
COVID-19	NCT04522986	Completed	Phase 1	6
ARDS, COVID-19	NCT04390139	Active	Phase 1, 2	30
ARDS, COVID-19	NCT04348461	Suspended	Phase 2	100
COVID-19	NCT04397796	Active	Phase 1	45
Post COVID-19 Syndrome	NCT05126563	Recruiting	Phase 2	80
COVID-19 Pneumonia, Multiple Organ Failure	NCT04392778	Completed	Phase 1, 2	30
COVID-19 Pneumonia	NCT04798716	Not recruiting	Phase 1, 2	55
ARDS	NCT05491681	Not recruiting	Phase 1	9
Post COVID-19 Syndrome	NCT04798066	No longer available		
COVID-19	NCT04467047	Unknown status	Phase 1	10

COVID-19 Pneumonia	NCT04398303	Unknown status	Phase 1, 2	70
COVID-19 Pneumonia	NCT04361942	Completed	Phase 2	24
COVID-19	NCT04437823	Unknown status	Phase 2	20
ARDS, COVID-19	NCT04537351	Completed	Phase 1, 2	14
ARDS, COVID-19	NCT04602442	Unknown status	Phase 2	90
COVID-19, ARDS Pneumonia	NCT05092724	Not recruiting	Not Applicable	20
COVID-19, Cytokine Storm	NCT05019287	Completed	Phase 1, 2	29
COVID-19 Pneumonia	NCT04491240	Completed	Phase 1, 2	30
ARDS, COVID-19	NCT04371393	Terminated	Phase 3	223
ARDS, COVID-19	NCT04269525	Unknown status	Phase 2	16
ARDS, COVID-19	NCT04524962	Recruiting	Phase 1, 2	30
SARS Pneumonia	NCT04299152	Not recruiting	Phase 2	20
ARDS, COVID-19	NCT04466098	Active	Phase 2	9
COVID-19, Acute Kidney Injury, Sepsis	NCT04445220	Active	Phase 1, 2	22
ARDS	NCT03042143	Recruiting	Phase 1, 2	129
Respiratory Tract Diseases	NCT04341610	Withdrawn	Phase 1, 2	0
ARDS, COVID-19	NCT04400032	Completed	Phase 1, 2	15
ARDS	NCT04333368	Completed	Phase 1, 2	47
Pulmonary Disorders	NCT04684602	Recruiting	Phase 1, 2	5000
COVID-19	NCT05348772	Not recruiting	Phase 1	18
COVID-19 Pneumonia	NCT05286255	Recruiting	Phase 1	10
IPF	NCT05468502	Recruiting	Phase 1	18
IPF	NCT01919827	Completed	Phase 1	17
IPF	NCT01385644	Completed	Phase 1	8
IPF	NCT05016817	Recruiting	Phase 1	20

Continued

Table 7 Representative MSC-based trials for respiratory disease treatment—cont'd

Disease	NCT number	Status	Phases	Enrollment
IPF	NCT02013700	Terminated	Phase 1	9
IPF	NCT02135380	Unknown status	Phase 1, 2	60
IPF	NCT02594839	Completed	Phase 1, 2	20
COPD	NCT04047810	Active	Phase 1	15
COPD	NCT04206007	Active	Phase 1	9
COPD	NCT04433104	Recruiting	Phase 1, 2	40
Pulmonary Disease, COPD, Chronic Bronchitis	NCT00683722	Completed	Phase 2	62
Pulmonary Disease, Asthma, COPD	NCT05147688	Recruiting	Phase 1	20
Severe COPD	NCT04018729	Recruiting	Phase 2, 3	34
COPD	NCT02645305	Unknown status	Phase 1, 2	20
Chronic Obstructive Lung Disease, COPD	NCT05227547	Recruiting	Not Applicable	186
Bronchopulmonary Dysplasia	NCT03558334	Unknown status	Phase 1	12
Bronchopulmonary Dysplasia	NCT03601416	Unknown status	Phase 2	57
Bronchopulmonary Dysplasia	NCT03645525	Recruiting	Phase 1, 2	180
Bronchopulmonary Dysplasia	NCT03873506	Unknown status	Phase 1	30
Bronchopulmonary Dysplasia	NCT04062136	Unknown status	Phase 1	10
Bronchopulmonary Dysplasia	NCT02443961	Active	Phase 1	10
Bronchopulmonary Dysplasia	NCT03378063	Recruiting	Early Phase 1	100
Bronchopulmonary Dysplasia	NCT03774537	Unknown status	Phase 1, 2	20
Bronchopulmonary Dysplasia	NCT03631420	Recruiting	Phase 1	9
Bronchopulmonary Dysplasia	NCT02381366	Completed	Phase 1, 2	12
Bronchopulmonary Dysplasia	NCT01207869	Unknown status	Phase 1	10

Bronchopulmonary Dysplasia	NCT01297205	Completed	Phase 1	9
Bronchopulmonary Dysplasia	NCT03857841	Terminated	Phase 1	3
Bronchopulmonary Dysplasia	NCT01828957	Completed	Phase 2	69
Bronchopulmonary Dysplasia	NCT04003857	Recruiting	Phase 2	60
Bronchopulmonary Dysplasia	NCT01897987	Completed	Not Applicable	62
Bronchopulmonary Dysplasia	NCT01632475	Active		9
Severe Bronchopulmonary Dysplasia	NCT03392467	Unknown status	Phase 2	60
Bronchopulmonary Dysplasia, Respiratory Tract Infections	NCT02023788	Completed		8
Bronchopulmonary Dysplasia	NCT04255147	Not recruiting	Phase 1	9

infertility has an immense impact upon the psychosocial wellbeing and quality of life of couples worldwide [79]. During the past few decades, tremendous headway has been achieved to improve the outcomes of patients with infertility, utilizing the systemic or local application of MSCs, and in particular, to those resulting from premature ovarian insufficiency (POI) and azoospermia [79].

Encouragingly, numerous studies have indicated the possibility of MSC-based treatment in a variety of reproductive disorders, including POI, premature ovarian failure (POF), polycystic ovary syndrome (PCOS), and intrauterine adhesion (IUA) [80]. For instance, Shareghi-Oskoue et al. reviewed the latest updates of UC-MSCs upon POF administration via increasing ovarian weight and the plasma E2 level, together with facilitating the generation of standard follicles [81]. Furthermore, Bahrehbar and colleagues took utilized a chemotherapy-induced POF mouse model and compared the ameliorative effect of ESC-MSCs and BM-MSCs upon restoring ovarian function via promoting ovarian secretion and promoting follicular development [82].

As an endometrial fibrosis disorder resulting from repeated uterus operations and endometrial traumas, IUA causes serious problems of endometrial regeneration and concomitant female infertility [83,84]. Researchers have summarized the latest literature related to IUA treatment and the resultant Asherman's syndrome using MSC-based strategies, including allogeneic or autologous MSCs, UC-MSC-derived conditioned medium, adipose MSC-derived exosomes, and injectable MSC-laden Matrigel microspheres [85–87]. For example, Liu et al. reported the strides made toward the effective management of endometrial injury in Asherman's syndrome rats by hyaluronic acid hydrogel/MSC-secretome composite [88].

As the most common pathogenic factor in anovulatory infertility in women, PCOS is a long-unresolved disorder that represents a critical threat to female reproduction [89,90]. Due to the deficiency of effective remedies for PCOS management, investigators in the field have turned to exploring the therapeutic role of MSCs and their derivations in ameliorating the clinical manifestations and female infertility [90,91]. For example, Zhao et al. recently verified the efficacy of UC-MSC-derived exosomes upon PCOS and polycystic ovary syndrome through inhibiting the NK-κB signaling-mediated inflammatory response of ovarian granulosa cells [92].

Taken together, state-of-the-art research has highlighted the great promise of MSCs and its derivations such as MSC-derived exosomes (MSC-Exo) and small extracellular vesicles (MSC-sEVs) for improving the pregnancy outcomes of infertile couples affected by conditions such as intrauterine adhesion, endometrial fibrosis, injured endometrium, thin endometrium, POI, POF, PCOS, polycystic ovary syndrome, and diminished ovarian reserve [80,93].

Due to the rising incidence and the high fatality rate, kidney diseases with short-term or long-term adverse outcomes have posed a threat to the patients. Worse still, the aforementioned kidney diseases are adequate to develop into functional organ failure [94,95]. At present, MSCs and the aforementioned derivations (e.g., MSC-Exo, MSC-sEVs) have become a research hotspot for targeted therapy of a variety of kidney diseases, such as atherosclerotic renovascular disease (ARVD), acute kidney injury (AKI), diabetic nephropathy (DN), and chronic kidney disease (CKD) [96,97]. For instance, Yu and colleagues recently reported the efficacy and engraftment of MSCs in interstitial cystitis/bladder pain syndrome administration by utilizing single cell transcriptomic analysis and intravital imaging technologies [98]. In particular, they verified the key characteristics of cyclin dependent kinase-1 (CDK1) and FOS in modulating the engraftment, migration, and antiinflammatory effect of multipotent MSCs [98]. Furthermore, Gao and colleagues demonstrated the effectiveness of hypoxic MSC-sEVs in ameliorating renal fibrosis of ischemia-reperfusion injury through restoring the fatty acid oxidation mediated by carnitine palmitoyl-transferase 1A (CPT1A) [99].

Recent progress in fundamental studies and clinical investigation has suggested the mode of action of MSCs and their derivations in urogenital disease management, including microenvironment optimization for cell survival, functional repair of damaged tissues, and suppression of inflammatory responses. Despite this, the detailed molecular mechanism of MSC-based cytotherapy remains uncertain; however, the therapeutic effect in the treatment or improvement of a diverse range of refractory reproductive and urinary disorders has been extensively investigated in clinical trials.

As shown by Table 8, considerable numbers of MSC-based trials for urogenital disease administration have been registered and conducted, including POI, POF, fallopian tube obstruction, male infertility, uterine scar, nonobstructive azoospermia, stress urinary incontinence, interstitial cystitis, and uterine scar, which collectively indicate the promising prospects of MSC-based remedies in the future. Of note, in consideration of the risk of allosensitization, Kim et al. recognized that autologous MSC-based transplantation was preferred over allogenic cytotherapy in patients with CKD, which indicated the variations in curative effect of MSCs in disease

Table 8 Representative MSC-based trials for urogenital disease treatment.

Conditions	NCT number	Status	Phases	Enrollment
Premature Ovarian Failure	NCT03816852	Suspended	Phase 2	12
Premature Ovarian Failure	NCT03069209	Unknown status	Phase 1, 2	50
Premature Ovarian Failure	NCT02062931	Unknown status	Phase 1, 2	60
Premature Ovarian Failure	NCT01742533	Unknown status	Phase 1, 2	40
Premature Ovarian Insufficiency	NCT05308342	Recruiting	Not Applicable	66
Primary Ovarian Insufficiency	NCT03877471	Active	Phase 1	28
Primary Ovarian Insufficiency	NCT03033277	Unknown status	Phase 1, 2	320
Primary Ovarian Insufficiency, Premature Ovarian Failure	NCT04675970	Active		86
Premature Ovarian Failure	NCT02603744	Unknown status	Phase 1, 2	9
Premature Ovarian Failure	NCT04815213	Recruiting	Phase 1	10
Premature Ovarian Failure	NCT02644447	Completed	Phase 1, 2	23
Premature Ovarian Failure	NCT02779374	Terminated	Not Applicable	10
Primary Ovarian Insufficiency	NCT05494723	Not recruiting	Phase 1	6
Premature Ovarian Failure	NCT02372474	Completed	Phase 1, 2	112
Premature Ovarian Failure	NCT03816852	Suspended	Phase 2	12
Premature Ovarian Insufficiency	NCT05308342	Recruiting	Not Applicable	66

Premature Ovarian Failure	NCT03069209	Unknown status	Phase 1, 2	50
Primary Ovarian Insufficiency	NCT03877471	Active	Phase 1	28
Premature Ovarian Failure	NCT02062931	Unknown status	Phase 1, 2	60
Premature Ovarian Failure	NCT01742533	Unknown status	Phase 1, 2	40
Primary Ovarian Insufficiency	NCT03033277	Unknown status	Phase 1, 2	320
Primary Ovarian Insufficiency, Premature Ovarian Failure	NCT04675970	Active		86
Premature Ovarian Failure	NCT02603744	Unknown status	Phase 1, 2	9
Premature Ovarian Failure	NCT04815213	Recruiting	Phase 1	10
Premature Ovarian Failure	NCT02644447	Completed	Phase 1, 2	23
Premature Ovarian Failure	NCT02779374	Terminated	Not Applicable	10
Primary Ovarian Insufficiency	NCT05494723	Not recruiting	Phase 1	6
Premature Ovarian Failure	NCT02372474	Completed	Phase 1, 2	112
Thin Endometrium, Intrauterine Adhesion	NCT03724617	Completed	Not Applicable	18
Infertility, Female Endometrium	NCT03592849	Completed	Not Applicable	20
Intrauterine Adhesion	NCT04232592	Unknown status	Phase 1	32
Thin Endometrium, Intrauterine Adhesion	NCT03724617	Completed	Not Applicable	18

Continued

Table 8 Representative MSC-based trials for urogenital disease treatment—cont'd

Conditions	NCT number	Status	Phases	Enrollment
Infertility, Intrauterine Adhesions	NCT02313415	Completed	Not Applicable	26
Chronic Endometritis, Uterine Synechiae, Fallopian Tube Obstruction	NCT04432467	Completed	Phase 1, 2	10
Nonobstructive Azoospermia	NCT02641769	Unknown status	Phase 1, 2	50
Male Infertility	NCT02025270	Unknown status	Phase 1, 2	100
Uterine Scar	NCT03181087	Withdrawn	Phase 1	0
Uterine Scar	NCT02968459	Withdrawn	Phase 2	0
Interstitial Cystitis	NCT04610359	Unknown status	Phase 1	3
Erectile Dysfunction, Peyronie's Disease, Interstitial Cystitis	NCT05147779	Recruiting	Phase 1	20
Stress Urinary Incontinence	NCT04446884	Completed	Phase 1, 2	10
Stress Urinary Incontinence	NCT02334878	Completed	Phase 3	50
Urinary Incontinence	NCT01804153	Unknown status	Phase 1, 2	10

treatment [100]. Collectively, the heterogeneous MSCs with immunomodulatory effects and secretory properties represent a novel therapeutic paradigm for urogenital diseases in preclinical and clinical practice.

Acknowledgment

This work was supported by grants from the National Natural Science Foundation of China (82260031), supported by Jiangxi Provincial Natural Science Foundation (20224BAB206077, 20212BAB216073), the Project Youth Fund, supported by Shandong Provincial Natural Science Foundation (ZR2020QC097), the Nonprofit Central Research Institute Fund of Chinese Academy of Medical Sciences (2019PT320005), and the 2021 Central-Guided Local Science and Technology Development Fund (ZYYDDFFZZJ-1), Gansu Provincial Hospital Intra-Hospital Research Fund Project (22GSSYB-6, 21GSSYB-8, 20GSSY5-2), The 2022 Master/Doctor/Postdoctoral program of NHC Key Laboratory of Diagnosis and Therapy of Gastrointestinal Tumor (NHCDP2022004, NHCDP2022008, NHCDP2022014), Key project funded by Department of Science and Technology of Shangrao City (2022A001), Double Thousand Plan of Jiangxi Province (2022 to LSZ).

Conflict of interest

The coauthors declare that there are no competing interests and all the coauthors consent to publish the data.

References

[1] Kanter J, Liem RI, Bernaudin F, Bolanos-Meade J, Fitzhugh CD, Hankins JS, Murad MH, Panepinto JA, Rondelli D, Shenoy S, et al. American Society of Hematology 2021 guidelines for sickle cell disease: stem cell transplantation. Blood Adv 2021;5(18):3668–89.

[2] Xu ZL, Huang XJ. Optimizing allogeneic grafts in hematopoietic stem cell transplantation. Stem Cells Transl Med 2021;10(Suppl 2):S41–7.

[3] Wu Y, Wang Z, Cao Y, Xu L, Li X, Liu P, Yan P, Liu Z, Zhao D, Wang J, et al. Cotransplantation of haploidentical hematopoietic and umbilical cord mesenchymal stem cells with a myeloablative regimen for refractory/relapsed hematologic malignancy. Ann Hematol 2013;92(12):1675–84.

[4] Li O, Tormin A, Sundberg B, Hyllner J, Le Blanc K, Scheding S. Human embryonic stem cell-derived mesenchymal stroma cells (hES-MSCs) engraft in vivo and support hematopoiesis without suppressing immune function: implications for off-the shelf ES-MSC therapies. PLoS One 2013;8(1)e55319.

[5] Hou H, Zhang L, Duan L, Liu Y, Han Z, Li Z, Cao X. Spatio-temporal metabolokinetics and efficacy of human placenta-derived mesenchymal stem/stromal cells on mice with refractory Crohn's-like enterocutaneous fistula. Stem Cell Rev Rep 2020;16(6):1292–304.

[6] Zhang L, Wang H, Liu C, Wu Q, Su P, Wu D, Guo J, Zhou W, Xu Y, Shi L, et al. MSX2 initiates and accelerates mesenchymal stem/stromal cell specification of hPSCs by regulating TWIST1 and PRAME. Stem Cell Rep 2018;11(2):497–513.

[7] Yao J, Chen N, Wang X, Zhang L, Huo J, Chi Y, Li Z, Han Z. Human supernumerary teeth-derived apical papillary stem cells possess preferable characteristics and efficacy on hepatic fibrosis in mice. Stem Cells Int 2020;2020:6489396.
[8] Ghavamzadeh A, Mirzania M, Kamalian N, Sedighi N, Azimi P. Hepatic iron overload and fibrosis in patients with beta thalassemia major after hematopoietic stem cell transplantation: a pilot study. Int J Hematol Oncol Stem Cell Res 2015;9(2):55–9.
[9] Mohamadnejad M, Alimoghaddam K, Mohyeddinbonab M. Phase 1 trial of autologous bone marrow mesenchymal stem cell transplantation in patients with decompensated liver cirrhosis. Arch Iran Med 2007;10(4):459–66.
[10] Mohamadnejad M, Alimoghaddam K, Mohyeddin-Bonab M, Bagheri M, Bashtar M, Ghanaati H, Baharvand H, Ghavamzadeh A, Malekzadeh R. Phase 1 trial of autologous bone marrow mesenchymal stem cell transplantation in patients with decompensated liver cirrhosis. Arch Iran Med 2007;10(4):459–66.
[11] Sanchez L, Gutierrez-Aranda I, Ligero G, Rubio R, Munoz-Lopez M, Garcia-Perez JL, Ramos V, Real PJ, Bueno C, Rodriguez R, et al. Enrichment of human ESC-derived multipotent mesenchymal stem cells with immunosuppressive and anti-inflammatory properties capable to protect against experimental inflammatory bowel disease. Stem Cells 2011;29(2):251–62.
[12] Wang X, Lazorchak AS, Song L, Li E, Zhang Z, Jiang B, Xu RH. Immune modulatory mesenchymal stem cells derived from human embryonic stem cells through a trophoblast-like stage. Stem Cells 2016;34(2):380–91.
[13] Liang L, Dong C, Chen X, Fang Z, Xu J, Liu M, Zhang X, Gu DS, Wang D, Du W, et al. Human umbilical cord mesenchymal stem cells ameliorate mice trinitrobenzene sulfonic acid (TNBS)-induced colitis. Cell Transplant 2011;20(9):1395–408.
[14] Hou H, Zhang L, Duan L, Liu Y, Han Z, Li Z, Cao X. Spatio-temporal metabolokinetics and efficacy of human placenta-derived mesenchymal stem/stromal cells on mice with refractory Crohn's-like enterocutaneous fistula. Stem Cell Rev Rep 2020;16:
1292–304.
[15] Guerreiro S, Privat AL, Bressac L, Toulorge D. CD38 in neurodegeneration and neuroinflammation. Cell 2020;9(2):471.
[16] Wyss-Coray T. Ageing, neurodegeneration and brain rejuvenation. Nature 2016;539 (7628):180–6.
[17] Kritsilis M, Rizou SV, Koutsoudaki PN, Evangelou K, Gorgoulis VG, Papadopoulos D. Ageing, cellular senescence and neurodegenerative disease. Int J Mol Sci 2018;19(10):2937.
[18] Andrzejewska A, Dabrowska S, Lukomska B, Janowski M. Mesenchymal stem cells for neurological disorders. Adv Sci 2021;8(7):2002944.
[19] Czarzasta J, Habich A, Siwek T, Czaplinski A, Maksymowicz W, Wojtkiewicz J. Stem cells for ALS: an overview of possible therapeutic approaches. Int J Dev Neurosci 2017;57:46–55.
[20] Mazzini L, Gelati M, Profico DC, Sgaravizzi G, Projetti Pensi M, Muzi G, Ricciolini C, Rota Nodari L, Carletti S, Giorgi C, et al. Human neural stem cell transplantation in ALS: initial results from a phase I trial. J Transl Med 2015;13:17.
[21] Sykova E, Rychmach P, Drahoradova I, Konradova S, Ruzickova K, Vorisek I, Forostyak S, Homola A, Bojar M. Transplantation of mesenchymal stromal cells in patients with amyotrophic lateral sclerosis: results of phase I/IIa clinical trial. Cell Transplant 2017;26(4):647–58.
[22] Reza-Zaldivar EE, Hernandez-Sapiens MA, Minjarez B, Gutierrez-Mercado YK, Marquez-Aguirre AL, Canales-Aguirre AA. Potential effects of MSC-derived exosomes in neuroplasticity in Alzheimer's disease. Front Cell Neurosci 2018;12:317.

[23] Ding R, Lin C, Wei S, Zhang N, Tang L, Lin Y, Chen Z, Xie T, Chen X, Feng Y, et al. Therapeutic benefits of mesenchymal stromal cells in a rat model of hemoglobin-induced hypertensive intracerebral hemorrhage. Mol Cell 2017;40(2):133–42.

[24] Wang X, Kimbrel EA, Ijichi K, Paul D, Lazorchak AS, Chu J, Kouris NA, Yavanian GJ, Lu SJ, Pachter JS, et al. Human ESC-derived MSCs outperform bone marrow MSCs in the treatment of an EAE model of multiple sclerosis. Stem Cell Rep 2014;3(1):115–30.

[25] Matsuura S, Shuvaev AN, Iizuka A, Nakamura K, Hirai H. Mesenchymal stem cells ameliorate cerebellar pathology in a mouse model of spinocerebellar ataxia type 1. Cerebellum 2014;13(3):323–30.

[26] Köhli P, Otto E, Jahn D, Reisener MJ, Appelt J, Rahmani A, Taheri N, Keller J, Pumberger M, Tsitsilonis S. Future perspectives in spinal cord repair: brain as saviour? TSCI with concurrent TBI: pathophysiological interaction and impact on MSC treatment. Cell 2021;10(11):2955.

[27] Correia JS, Neves-Carvalho A, Mendes-Pinheiro B, Pires J, Teixeira FG, Lima R, Monteiro S, Silva NA, Soares-Cunha C, Serra SC, Duarte-Silva S, Teixeira-Castro A, Salgado AJ, Maciel P. Preclinical assessment of mesenchymal-stem-cell-based therapies in spinocerebellar ataxia type 3. Biomedicine 2021;9(12):1754.

[28] Cho J, D'Antuono M, Glicksman M, Wang J, Jonklaas J. A review of clinical trials: mesenchymal stem cell transplant therapy in type 1 and type 2 diabetes mellitus. Am J Stem Cells 2018;7(4):82–93.

[29] Aydin S, Sagrac D, Sahin F. Differentiation potential of mesenchymal stem cells into pancreatic beta-cells. Adv Exp Med Biol 2020;1247:135–56.

[30] Lv W, Graves DT, He L, Shi Y, Deng X, Zhao Y, Dong X, Ren Y, Liu X, Xiao E, et al. Depletion of the diabetic gut microbiota resistance enhances stem cells therapy in type 1 diabetes mellitus. Theranostics 2020;10(14):6500–16.

[31] Jiang R, Han Z, Zhuo G, Qu X, Li X, Wang X, Shao Y, Yang S, Han ZC. Transplantation of placenta-derived mesenchymal stem cells in type 2 diabetes: a pilot study. Front Med 2011;5(1):94–100.

[32] Fiori A, Terlizzi V, Kremer H, Gebauer J, Hammes HP, Harmsen MC, Bieback K. Mesenchymal stromal/stem cells as potential therapy in diabetic retinopathy. Immunobiology 2018;223(12):729–43.

[33] Zhao K, Liu J, Dong G, Xia H, Wang P, Xiao X, Chen Z. Preliminary research on the effects and mechanisms of umbilical cordderived mesenchymal stem cells in streptozotocininduced diabetic retinopathy. Int J Mol Med 2020;46(2):849–58.

[34] Ramasamy SK, Kusumbe AP, Wang L, Adams RH. Endothelial notch activity promotes angiogenesis and osteogenesis in bone. Nature 2014;507(7492):376–80.

[35] Yang X, Xu S, Chen X, He D, Ke X, Zhang L, Yang G, Liu A, Mou X, Xia W, et al. Intra-bone marrow injection of trace elements co-doped calcium phosphate microparticles for the treatment of osteoporotic rat. J Biomed Mater Res A 2017;105(5):1422–32.

[36] Zak A, Tvrzicka E, Vecka M, Jachymova M, Duffkova L, Stankova B, Vavrova L, Kodydkova J, Zeman M. Severity of metabolic syndrome unfavorably influences oxidative stress and fatty acid metabolism in men. Tohoku J Exp Med 2007;212(4):359–71.

[37] Zhao Q, Zhang L, Wei Y, Yu H, Zou L, Huo J, Yang H, Song B, Wei T, Wu D, et al. Systematic comparison of hUC-MSCs at various passages reveals the variations of signatures and therapeutic effect on acute graft-versus-host disease. Stem Cell Res Ther 2019;10(1):354.

[38] Fillatreau S, Manfroi B, Dorner T. Toll-like receptor signalling in B cells during systemic lupus erythematosus. Nat Rev Rheumatol 2021;17(2):98–108.

[39] Peltzer J, Aletti M, Frescaline N, Busson E, Lataillade JJ, Martinaud C. Mesenchymal stromal cells based therapy in systemic sclerosis: rational and challenges. Front Immunol 2013;2018:9.
[40] Li H, Tian Y, Xie L, Liu X, Huang Z, Su W. Mesenchymal stem cells in allergic diseases: current status. Allergol Int 2020;69(1):35–45.
[41] Christopeit M, Schendel M, Foll J, Muller LP, Keysser G, Behre G. Marked improvement of severe progressive systemic sclerosis after transplantation of mesenchymal stem cells from an allogeneic haploidentical-related donor mediated by ligation of CD137L. Leukemia 2008;22(5):1062–4.
[42] Thiel A, Yavanian G, Nastke MD, Morales P, Kouris NA, Kimbrel EA, Lanza R. Human embryonic stem cell-derived mesenchymal cells preserve kidney function and extend lifespan in NZB/W F1 mouse model of lupus nephritis. Sci Rep 2015;5:17685.
[43] Wen L, Labopin M, Badoglio M, Wang D, Sun L, Farge-Bancel D. Prognostic factors for clinical response in systemic lupus erythematosus patients treated by allogeneic mesenchymal stem cells. Stem Cells Int 2019;2019:7061408.
[44] Granel B, Daumas A, Jouve E, Harle JR, Nguyen PS, Chabannon C, Colavolpe N, Reynier JC, Truillet R, Mallet S, et al. Safety, tolerability and potential efficacy of injection of autologous adipose-derived stromal vascular fraction in the fingers of patients with systemic sclerosis: an open-label phase I trial. Ann Rheum Dis 2015;74(12):2175–82.
[45] Buzaboon N, Alshammary S. Clinical applicability of adult human mesenchymal stem cell therapy in the treatment of knee osteoarthritis. Stem Cells Cloning 2020;13: 117–36.
[46] Glyn-Jones S, Palmer AJ, Agricola R, Price AJ, Vincent TL, Weinans H, Carr AJ. Osteoarthritis. Lancet 2015;386(9991):376–87.
[47] Jo CH, Lee YG, Shin WH, Kim H, Chai JW, Jeong EC, Kim JE, Shim H, Shin JS, Shin IS, et al. Intra-articular injection of mesenchymal stem cells for the treatment of osteoarthritis of the knee: a proof-of-concept clinical trial. Stem Cells 2014;32(5): 1254–66.
[48] Zhang L, Wei Y, Chi Y, Liu D, Yang S, Han Z, Li Z. Two-step generation of mesenchymal stem/stromal cells from human pluripotent stem cells with reinforced efficacy upon osteoarthritis rabbits by HA hydrogel. Cell Biosci 2021;11(1):6.
[49] Lian Q, Zhang Y, Zhang J, Zhang HK, Wu X, Zhang Y, Lam FF, Kang S, Xia JC, Lai WH, et al. Functional mesenchymal stem cells derived from human induced pluripotent stem cells attenuate limb ischemia in mice. Circulation 2010;121(9):1113–23.
[50] Vangsness Jr. CT, Farr 2nd J, Boyd J, Dellaero DT, Mills CR, LeRoux-Williams M. Adult human mesenchymal stem cells delivered via intra-articular injection to the knee following partial medial meniscectomy: a randomized, double-blind, controlled study. J Bone Joint Surg Am 2014;96(2):90–8.
[51] Leng Z, Zhu R, Hou W, Feng Y, Yang Y, Han Q, Shan G, Meng F, Du D, Wang S, et al. Transplantation of ACE2(−) mesenchymal stem cells improves the outcome of patients with COVID-19 pneumonia. Aging Dis 2020;11(2):216–28.
[52] Yen BL, Yen ML, Wang LT, Liu KJ, Sytwu HK. Current status of mesenchymal stem cell therapy for immune/inflammatory lung disorders: gleaning insights for possible use in COVID-19. Stem Cells Transl Med 2020;9(10):1163–73.
[53] Meng F, Xu R, Wang S, Xu Z, Zhang C, Li Y, Yang T, Shi L, Fu J, Jiang T, et al. Human umbilical cord-derived mesenchymal stem cell therapy in patients with COVID-19: a phase 1 clinical trial. Signal Transduct Target Ther 2020;5(1):172.
[54] Shi L, Huang H, Lu X, Yan X, Jiang X, Xu R, Wang S, Zhang C, Yuan X, Xu Z, et al. Effect of human umbilical cord-derived mesenchymal stem cells on lung damage in severe COVID-19 patients: a randomized, double-blind, placebo-controlled phase 2 trial. Signal Transduct Target Ther 2021;6(1):58.

[55] Monsel A, Zhu YG, Gennai S, Hao Q, Hu S, Rouby JJ, Rosenzwajg M, Matthay MA, Lee JW. Therapeutic effects of human mesenchymal stem cell-derived microvesicles in severe pneumonia in mice. Am J Respir Crit Care Med 2015;192(3):324–36.
[56] Zhang LS, Yu Y, Yu H, Han ZC. Therapeutic prospects of mesenchymal stem/stromal cells in COVID-19 associated pulmonary diseases: from bench to bedside. World J Stem Cells 2021;13(8):1058–71.
[57] Bouvier NM, Lowen AC. Animal models for influenza virus pathogenesis and transmission. Viruses 2010;2(8):1530–63.
[58] Fukushi M, Ito T, Oka T, Kitazawa T, Miyoshi-Akiyama T, Kirikae T, Yamashita M, Kudo K. Serial histopathological examination of the lungs of mice infected with influenza A virus PR8 strain. PLoS One 2011;6(6)e21207.
[59] Khatri M, Richardson LA, Meulia T. Mesenchymal stem cell-derived extracellular vesicles attenuate influenza virus-induced acute lung injury in a pig model. Stem Cell Res Ther 2018;9(1):17.
[60] Sun J, Han ZB, Liao W, Yang SG, Yang Z, Yu J, Meng L, Wu R, Han ZC. Intrapulmonary delivery of human umbilical cord mesenchymal stem cells attenuates acute lung injury by expanding CD4+CD25+ Forkhead Boxp3 (FOXP3)+ regulatory T cells and balancing anti- and pro-inflammatory factors. Cell Physiol Biochem 2011;27(5):
587–96.
[61] Rojas M, Cardenes N, Kocyildirim E, Tedrow JR, Caceres E, Deans R, Ting A, Bermudez C. Human adult bone marrow-derived stem cells decrease severity of lipopolysaccharide-induced acute respiratory distress syndrome in sheep. Stem Cell Res Ther 2014;5(2):42.
[62] Hao Q, Zhu YG, Monsel A, Gennai S, Lee T, Xu F, Lee JW. Study of bone marrow and embryonic stem cell-derived human mesenchymal stem cells for treatment of *Escherichia coli* endotoxin-induced acute lung injury in mice. Stem Cells Transl Med 2015;4(7):832–40.
[63] Horie S, Masterson C, Devaney J, Laffey JG. Stem cell therapy for acute respiratory distress syndrome: a promising future? Curr Opin Crit Care 2016;22(1):14–20.
[64] D'Alessio FR. Mouse models of acute lung injury and ARDS. Methods Mol Biol 2018;1809:341–50.
[65] Matthay MA, Calfee CS, Zhuo H, Thompson BT, Wilson JG, Levitt JE, Rogers AJ, Gotts JE, Wiener-Kronish JP, Bajwa EK, et al. Treatment with allogeneic mesenchymal stromal cells for moderate to severe acute respiratory distress syndrome (START study): a randomised phase 2a safety trial. Lancet Respir Med 2019;7(2):154–62.
[66] Aitong W, Leisheng Z, Hao Y. Visualized analyses of investigations upon mesenchymal stem/stromal cell-based cytotherapy and underlying mechanisms for COVID-19 associated ARDS. Curr Stem Cell Res Ther 2021;17:2–12.
[67] Ni K, Liu M, Zheng J, Wen L, Chen Q, Xiang Z, Lam KT, Liu Y, Chan GC, Lau YL, et al. PD-1/PD-L1 pathway mediates the alleviation of pulmonary fibrosis by human mesenchymal stem cells in humanized mice. Am J Respir Cell Mol Biol 2018;58(6): 684–95.
[68] Bouros D. Pirfenidone for idiopathic pulmonary fibrosis. Lancet 2011;377 (9779):1727–9.
[69] Chambers DC, Enever D, Ilic N, Sparks L, Whitelaw K, Ayres J, Yerkovich ST, Khalil D, Atkinson KM, Hopkins PM. A phase 1b study of placenta-derived mesenchymal stromal cells in patients with idiopathic pulmonary fibrosis. Respirology 2014; 19(7):1013–8.
[70] Srour N, Thebaud B. Mesenchymal stromal cells in animal bleomycin pulmonary fibrosis models: a systematic review. Stem Cells Transl Med 2015;4(12):1500–10.
[71] Chen S, Cui G, Peng C, Lavin MF, Sun X, Zhang E, Yang Y, Guan Y, Du Z, Shao H. Transplantation of adipose-derived mesenchymal stem cells attenuates pulmonary fibrosis of silicosis via anti-inflammatory and anti-apoptosis effects in rats. Stem Cell Res Ther 2018;9(1):110.

[72] Willis GR, Fernandez-Gonzalez A, Anastas J, Vitali SH, Liu X, Ericsson M, Kwong A, Mitsialis SA, Kourembanas S. Mesenchymal stromal cell exosomes ameliorate experimental bronchopulmonary dysplasia and restore lung function through macrophage immunomodulation. Am J Respir Crit Care Med 2018;197(1):104–16.
[73] Sun YQ, Deng MX, He J, Zeng QX, Wen W, Wong DS, Tse HF, Xu G, Lian Q, Shi J, et al. Human pluripotent stem cell-derived mesenchymal stem cells prevent allergic airway inflammation in mice. Stem Cells 2012;30(12):2692–9.
[74] Abreu SC, Antunes MA, Mendonca L, Branco VC, de Melo EB, Olsen PC, Diaz BL, Weiss DJ, Paredes BD, Xisto DG, et al. Effects of bone marrow mononuclear cells from healthy or ovalbumin-induced lung inflammation donors on recipient allergic asthma mice. Stem Cell Res Ther 2014;5(5):108.
[75] Shin JW, Ryu S, Ham J, Jung K, Lee S, Chung DH, Kang HR, Kim HY. Mesenchymal stem cells suppress severe asthma by directly regulating Th2 cells and type 2 innate lymphoid cells. Mol Cell 2021;44(8):580–90.
[76] Weiss DJ, Casaburi R, Flannery R, LeRoux-Williams M, Tashkin DP. A placebo-controlled, randomized trial of mesenchymal stem cells in COPD. Chest 2013;143(6):1590–8.
[77] Hoang DM, Pham PT, Bach TQ, Ngo ATL, Nguyen QT, Phan TTK, Nguyen GH, Le PTT, Hoang VT, Forsyth NR, Heke M, Nguyen LT. Stem cell-based therapy for human diseases. Signal Transduct Target Ther 2022;7(1):272. https://doi.org/10.1038/s41392-022-01134-4.
[78] Yin JQ, Zhu J, Ankrum JA. Manufacturing of primed mesenchymal stromal cells for therapy. Nat Biomed Eng 2019;3(2):90–104.
[79] Saha S, Roy P, Corbitt C, Kakar SS. Application of stem cell therapy for infertility. Cell 2021;10(7):1613. https://doi.org/10.3390/cells10071613.
[80] Liao Z, Liu C, Wang L, Sui C, Zhang H. Therapeutic role of mesenchymal stem cell-derived extracellular vesicles in female reproductive diseases. Front Endocrinol 2021;12:
665645.
[81] Shareghi-Oskoue O, Aghebati-Maleki L, Yousefi M. Transplantation of human umbilical cord mesenchymal stem cells to treat premature ovarian failure. Stem Cell Res Ther 2021;12(1):454. https://doi.org/10.1186/s13287-021-02529-w.
[82] Bahrehbar K, Rezazadeh Valojerdi M, Esfandiari F, Fathi R, Hassani SN, Baharvand H. Human embryonic stem cell-derived mesenchymal stem cells improved premature ovarian failure. World J Stem Cells 2020;12(8):857–78.
[83] Xin L, Lin X, Zhou F, Li C, Wang X, Yu H, Pan Y, Fei H, Ma L, Zhang S. A scaffold laden with mesenchymal stem cell-derived exosomes for promoting endometrium regeneration and fertility restoration through macrophage immunomodulation. Acta Biomater 2020;(113):252–66.
[84] Chen JM, Huang QY, Zhao YX, Chen WH, Lin S, Shi QY. The latest developments in immunomodulation of mesenchymal stem cells in the treatment of intrauterine adhesions, both allogeneic and autologous. Front Immunol 2021;12:785717.
[85] Xu B, Cao Y, Zheng Z, Galan EA, Hu Z, Ge J, Xing X, Ma S. Injectable mesenchymal stem cell-laden matrigel microspheres for endometrium repair and regeneration. Adv Biol 2021;5(8)e2000202.
[86] Zhao S, Qi W, Zheng J, Tian Y, Qi X, Kong D, Zhang J, Huang X. Exosomes derived from adipose mesenchymal stem cells restore functional endometrium in a rat model of intrauterine adhesions. Reprod Sci 2020;27(6):1266–75.
[87] Wei X, Liu F, Zhang S, Xu X, Li J, Wang Q, Cai J, Wang S. Human umbilical cord mesenchymal stem cell-derived conditioned medium promotes human endometrial cell proliferation through Wnt/β-catenin signaling. Biomed Res Int 2022;2022:8796093.

[88] Liu F, Hu S, Yang H, Li Z, Huang K, Su T, Wang S, Cheng K. Hyaluronic acid hydrogel integrated with mesenchymal stem cell-secretome to treat endometrial injury in a rat model of Asherman's syndrome. Adv Healthc Mater 2019;8(14)e1900411.
[89] Xie Q, Xiong X, Xiao N, He K, Chen M, Peng J, Su X, Mei H, Dai Y, Wei D, Lin G, Cheng L. Mesenchymal stem cells alleviate DHEA-induced polycystic ovary syndrome (PCOS) by inhibiting inflammation in mice. Stem Cells Int 2019;2019: 9782373.
[90] Liao Z, Liu C, Wang L, Sui C, Zhang H. Therapeutic role of mesenchymal stem cell-derived extracellular vesicles in female reproductive diseases. Front Endocrinol 2021;12:665645.
[91] Kalhori Z, Azadbakht M, Soleimani Mehranjani M, Shariatzadeh MA. Improvement of the folliculogenesis by transplantation of bone marrow mesenchymal stromal cells in mice with induced polycystic ovary syndrome. Cytotherapy 2018;20(12):1445–58.
[92] Zhao Y, Pan S, Wu X. Human umbilical cord mesenchymal stem cell-derived exosomes inhibit ovarian granulosa cells inflammatory response through inhibition of NF-κB signaling in polycystic ovary syndrome. J Reprod Immunol 2022;152:103638.
[93] Izadi M, Rezvani ME, Aliabadi A, Karimi M, Aflatoonian B. Mesenchymal stem cells-derived exosomes as a promising new approach for the treatment of infertility caused by polycystic ovary syndrome. Front Pharmacol 2022;(13):1021581. https://doi.org/10.3389/fphar.2022.1021581.
[94] Yun CW, Lee SH. Potential and therapeutic efficacy of cell-based therapy using mesenchymal stem cells for acute/chronic kidney disease. Int J Mol Sci 2019;20(7):1619.
[95] Prowle JR, Forni LG, Bell M, Chew MS, Edwards M, Grams ME, Grocott MPW, Liu KD, McIlroy D, Murray PT, Ostermann M, Zarbock A, Bagshaw SM, Bartz R, Bell S, Bihorac A, Gan TJ, Hobson CE, Joannidis M, Koyner JL, Levett DZH, Mehta RL, Miller TE, Mythen MG, Nadim MK, Pearse RM, Rimmele T, Ronco C, Shaw AD, Kellum JA. Postoperative acute kidney injury in adult non-cardiac surgery: joint consensus report of the acute disease quality initiative and PeriOperative quality initiative. Nat Rev Nephrol 2021;17(9):605–18.
[96] Huang Y, Yang L. Mesenchymal stem cells and extracellular vesicles in therapy against kidney diseases. Stem Cell Res Ther 2021;12(1):219.
[97] Uchida N, Kumagai N, Kondo Y. Application of muse cell therapy for kidney diseases. Adv Exp Med Biol 2018;1103:199–218.
[98] Yu HY, Lee S, Ju H, Kim Y, Shin JH, Yun H, Ryu CM, Heo J, Lim J, Song S, Lee S, Hong KS, Chung HM, Kim JK, Choo MS, Shin DM. Intravital imaging and single cell transcriptomic analysis for engraftment of mesenchymal stem cells in an animal model of interstitial cystitis/bladder pain syndrome. Biomaterials 2022;280:121277.
[99] Gao Z, Zhang C, Peng F, Chen Q, Zhao Y, Chen L, Wang X, Chen X. Hypoxic mesenchymal stem cell-derived extracellular vesicles ameliorate renal fibrosis after ischemia-reperfusion injure by restoring CPT1A mediated fatty acid oxidation. Stem Cell Res Ther 2022;13(1):191.
[100] Kim H, Yu MR, Lee H, Kwon SH, Jeon JS, Han DC, Noh H. Metformin inhibits chronic kidney disease-induced DNA damage and senescence of mesenchymal stem cells. Aging Cell 2021;20(2)e13317https://doi.org/10.1111/acel.13317 [Epub 2021 Feb 1].

CHAPTER 3

Biomaterials and MSCs composites in regenerative medicine

Chao Liu[a], Changsheng Shao[a], Leisheng Zhang[a,b,c,d,e,f], and Qing Huang[a]

[a]Key Laboratory of Radiation Technology and Biophysics, Hefei Institute of Physical Science, Chinese Academy of Sciences, Hefei, China
[b]Key Laboratory of Molecular Diagnostics and Precision Medicine for Surgical Oncology in Gansu Province & NHC Key Laboratory of Diagnosis and Therapy of Gastrointestinal Tumor, Gansu Provincial Hospital, Lanzhou, China
[c]School of Medicine, Nankai University, Tianjin, China
[d]National Engineering Technology Center for Cell Products & Jiangxi Research Center of Stem Cell Engineering, Jiangxi Health-Biotech Stem Cell Technology Co., Ltd., Shangrao, China
[e]Institute of Stem Cells, Health-Biotech (Tianjin) Stem Cell Research Institute Co., Ltd., Tianjin, China
[f]State Key Laboratory of Experimental Hematology & National Clinical Research Center for Blood Disease, Institute of Hematology & Blood Diseases Hospital, Chinese Academy of Medical Sciences & Peking Union Medical Colleague, Tianjin, China

3.1 Introduction to the types and characteristics of stem cells

Stem cells exist in embryos and adult tissues, and are a kind of undifferentiated cell with high self-renewal, proliferation, and multidirectional differentiation potential. So far, stem cells have been found in adult bone marrow, cartilage, blood, nerve, muscle, fat, skin, limbus, liver, pancreatic islets, and other tissues and organs. Stem cells are currently regarded as the most ideal seed cells for tissue engineering. Through tissue engineering technology, stem cells can theoretically be induced, differentiated, and cultured under artificial conditions into any human cell, tissue, or organ. In vivo transplantation of successfully cultured tissues and organs can perfectly repair or replace defective tissues and organs. According to the developmental process, stem cells can be divided into embryonic stem cells (ESCs) and adult stem cells (ASCs). The former are derived from embryonic or fetal tissue and can be differentiated into various cells of the three germ layers; the latter are derived from postnatal organs or adult human tissue, and can produce various cells in specific tissues. The adult stem cells discovered so far mainly include hematopoietic stem cells, bone marrow stromal stem cells, neural

Mesenchymal Stem Cells
https://doi.org/10.1016/B978-0-323-95346-7.00004-X

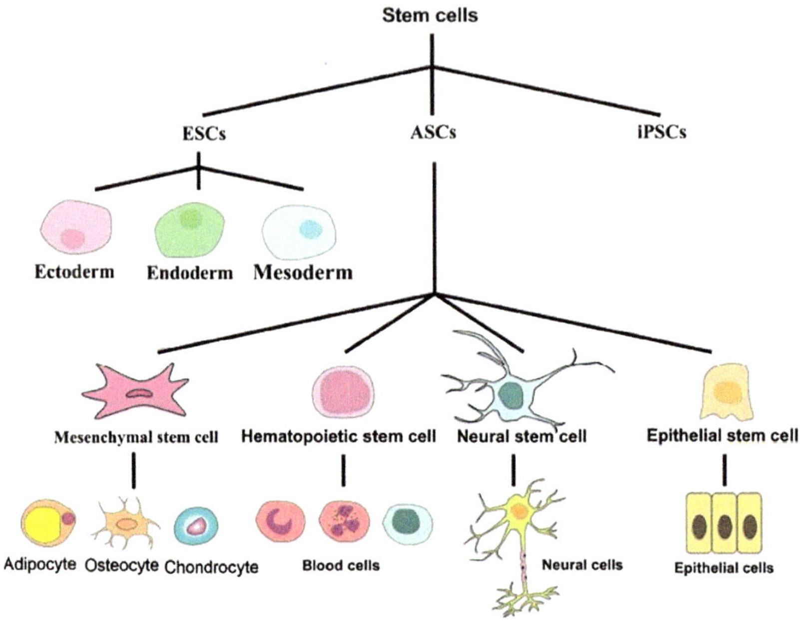

Fig. 1 Stem cell types and their differentiation lineage map: stem cells are divided into embryonic stem cells (ESCs), adult stem cells (ASCs), and induced pluripotent stem cells (iPSCs).

stem cells, liver stem cells, skin stem cells, intestinal epithelial stem cells, and adipose stem cells (Fig. 1).

Adult stem cells are very valuable research objects, and their advantage over embryonic stem cells is that they have superior practical performance in clinical applications. Stem cells can be used for the treatment of many diseases or injuries. Major breakthroughs that may occur in the treatment of diseases include skin attachment regeneration, degenerative disease treatment, cardiovascular disease treatment, etc. At present, it has been preliminarily observed in clinical applications that bone marrow stem cells have certain effects under specific conditions. These cells may develop into sweat glands, sebaceous glands, and hair follicles. Stem cells have been considered as the most ideal seed cells for tissue engineering and regenerative medicine. Generally, stem cells of specific types (e.g., human pluripotent stem cells including human embryonic stem cells and human induced pluripotent stem cells) can theoretically be programmed into various human cell types. Transplantation of the 3D tissues and organs generated from stem cells and biomaterials can perfectly repair or replace defective tissues and organs. ESCs and iPSCs (induced pluripotent stem cells) are pluripotent stem cells with high differentiation potential and almost unlimited self-renewal capacity. Mesenchymal stem cells (MSCs) are a type of fibroblast-like cell first discovered by Friedenstein et al., and offer tissue differentiation potential. ASCs in vivo originate from various tissues, such as bone marrow, adipose tissue, amniotic fluid, dental pulp, heart, and liver. Compared with the other two groups (ESCs and iPSCs), the differentiation capacity of ASCs is lower. Mesenchymal stem cells (MSCs) are a group of ASCs that can differentiate into different cell lines.

3.2 Introduction to biomaterials and tissue engineering

Biomaterial is an engineered material that is used in contact with biological tissue to evaluate, treat, or replace any tissue or organ of the organism, and to enhance and restore its function. Biomaterials are mainly used in the medical field, wherein it is processed and sterilized before being used. In recent years, researchers have used modified biomaterials to develop artificial organs to replace and repair damaged tissues and organs in the body, exert various physiological functions, and obtain good therapeutic effects; however, there are still certain problems. The current biomaterial technology fails to fully meet people's requirements, such as material and tissue compatibility, inflammatory response or tissue necrosis, immune response, etc.

Biomaterials can usually be divided into natural polymer biomaterials (regenerated fibers, collagen, hyaluronic acid, chitin), synthetic polymer biomaterials (silicone rubber, polyurethane and its block copolymers, polyester, nylon, polyacrylonitrile, polyolefins), medical metal materials (stainless steel, titanium and titanium alloys, titanium-nickel memory alloys), inorganic biomedical materials (carbon materials, bioactive ceramics, glass materials), hybrid biomaterials (referring to natural materials from living bodies and hybridization of synthetic materials, such as cross-linked hybridization of collagen and polyvinyl alcohol, etc.), and composite biomaterials (plastics reinforced with carbon fibers, bioceramics, glass reinforced with carbon fibers or glass fibers), etc.

The short-term implantation effect of some biomaterials is good, but the long-term effect is not good, and even leads to the failure of the implantation. These effects are due to the chemical structure and physical properties of biomaterials being very different from those of human tissues and organs; it is, therefore, necessary to use the recipient's own cells as biological substitute materials to repair defective tissues and organs. From this necessity, tissue engineering was born. Its core is to build a three-dimensional complex of cells and biological materials, which promotes the proliferation and division of seed cells. Therewith, the further development of biological materials in combination with diverse stem cells in the history of tissue engineering and regenerative medicine is inevitable and promising.

The definition of tissue engineering was clarified at the National Science Foundation (NSF) Expert Symposium held in Lake Tahoe, California in 1987 [1]:

> *Tissue engineering is the application of principles and methods of engineering and life sciences to fundamentally understand the relationship between normal and pathological mammalian tissue and structure-function, and to study biological substitutes to restore, maintain and improve tissue biological substitutes.*

3.3 Construction and application of different biomaterials/MSC composites

In past decades, MSCs have been widely used in the biological field, especially in tissue engineering. These cells are used to differentiate into heart cells and nerve cells, and can be used to repair heart and nerve damage. However, the efficiency and success of MSCs differentiation into these cells is low. On the other hand, undifferentiated implanted stem cells increase the risk of teratoma. Therefore, it is important to apply strategies to enhance differentiation and direct differentiation into specific cell lines.

As mentioned above, a three-dimensional complex of seed cells is constructed by using biological materials; that is, the expanded cells are mixed with biological materials with good biocompatibility, degradability, and absorbability in a certain proportion, so that the cells can adhere to the biological materials, forming cell-material complexes. Using the excellent physical and chemical properties of various biological materials to regulate the proliferation and differentiation of seed cells, and at the same time play their roles in drug delivery, electrical conductivity, targeting, and bionics, it is very important to give maximum attention to the role of materials science and life science in living organs. It is of great significance in reconstruction and disease treatment.

3.3.1 Scaffold materials and MSCs

In the reconstruction of tissues and organs, the three-dimensional structures that seed cells rely on for survival and attachment before regenerating tissues are called scaffolds; these are mainly biodegradable and absorbable biomaterials, called biomaterial scaffolds, and can hold cells in place. Consequently, one of the most critical issues in tissue engineering is to achieve specific physical, mechanical, and biological properties of scaffolds, because scaffolds provide a substrate for cell growth, proliferation, and support for the formation of new tissues. The biological tissue engineering scaffold material, we know, is a material that can be combined with the living cells of the tissue, which can be implanted into different tissues of the organism, and can replace the function of the tissue according to the specific situation. Therefore, in order to proliferate and differentiate seed cells, it is necessary to provide a cell scaffold composed of biological materials. The scaffold material is equivalent to an artificial extracellular matrix on which cells can adhere and proliferate.

The most basic feature of tissue engineering scaffolds is the direct binding to living cells, such as hydroxyapatite (HAP) directly binding to osteoblasts, polylactic acid (PLA) and polyglycolic acid (PGA) meshes that bind to chondrocytes, nylon mesh binding to hepatocytes, etc. Following on from this, an exploration of bioscaffold materials, biomaterials, and fabrication techniques plays a key role in tissue engineering. Materials to be used in tissue engineering must be engineered to stimulate specific cells at the molecular level, making them responsive. They should cause specific interactions between cells that directly affect cell adhesion, proliferation, differentiation, and extracellular matrix production and tissue generation. In addition, the combination with biological systems is also a basic feature of tissue engineering scaffold materials, such as the combination of tissue-engineered cartilage, bone, tendon, tooth, lung, liver, kidney, and other tissues directly implanted into the body, and the body itself. In addition to meeting various physical and chemical property requirements, tissue engineering scaffold materials must have cytocompatibility and histocompatibility (collectively referred to as biocompatibility), without exception.

Scaffolding materials for engineering bone and soft tissue include synthetic and natural calcium-phosphorus materials, and synthetic and natural polymers. Scaffolding materials for tissue engineering in vitro and implant materials for promoting regeneration in vivo need to have the necessary microstructure and chemical composition to accommodate mesenchymal cells and promote their function. From this point of view, the microporous structure of the material is compulsory. The required pore size, porosity, and micropore distribution and orientation should vary with different tissues. The chemical composition of the scaffold material is also important because it has an effect on cell attachment and immersion in cell phenotypic expression; and since our aim is to regenerate the original tissue, the material should be degradable, with the rate of degradation depending on the new tissue, the rate of formation, and the time for tissue remodeling at the implant site. Of course, the impact of the microparticles generated by the degradation of the scaffold material on the host and the regenerated tissue should also be considered. Finally, the mechanical properties of scaffolds for tissue engineering are also important, providing temporary support for in vivo load-bearing during tissue regeneration and resistance to the contractile forces generated by the seeded cells in the material before implantation and during cell filtration.

3.3.2 Common scaffold materials

Tissue engineering scaffold materials are designed according to the functions of materials used in different human tissues and specific substitute tissues. Due to the particularity of human tissue structure and the different functions of different tissues, many types of tissue engineering scaffold material have been designed; however, the most widely used tissue engineering scaffold matrix materials at present are natural materials, synthetic polymer degradable materials, bioceramic materials and their composite materials, as well as nano-tissue engineering materials, which have become research hotspots.

Natural high molecular polymers include collagen, fibrin, etc. These natural high polymer materials have good biocompatibility and have cell recognition signals (such as RGD sequence), which are beneficial to cell adhesion, proliferation, and differentiation. Natural biomaterials have many advantages over synthetic materials, such as abundant sources, low cost, and good biocompatibility. Some materials also have a natural pore structure system. At present, the most widely used include collagen (type I collagen), gelatin, chitin, chitosan, and natural coral and its derivatives, among which, collagen and chitin derivatives are the most widely used. The advantages of natural materials are that they contain substances (such as specific proteins) that facilitate the adsorption of cells or maintain different functions, and have good biocompatibility and cell affinity. However, the disadvantage is that the quality is affected by the place of origin, source of raw materials, etc., so the repeatability is poor and mass production is not possible. In addition, some natural macromolecules have poor strength and processing performance, and some are very expensive, making it difficult to use them directly as cell scaffolds. At the same time, the problem of xenotransplantation and the viral genes carried by xenogeneic organisms may bring unpredictable limitations in the application of such materials.

At present, the most widely used synthetic degradable polymer materials are polyglycolic acid (PGA), polylactic acid (PLA), and various copolymers between them. The main structural forms of PLA, PGA, and their copolymers used in tissue engineering are fibrous scaffolds, porous foams, and tubular structures. Due to the inherent shortcomings of poor mechanical properties, and the toxicity and side effects of degrading products, polymer materials cannot be widely used in tissue engineering, whereas degradable inorganic porous bioceramic materials have good biocompatibility and mechanical strength. At the same time, bioceramic materials have the degradability of polymer scaffold materials, providing broad application prospects in tissue engineering research fields such as cartilage and bone. Bioceramic materials are used as cell carrier scaffolds for bone tissue engineering. The effect is certain, but its shortcomings are also obvious. Firstly, it is brittle and not flexible enough; secondly, it is difficult to degrade in vivo, which affects the ingrowth and later reconstruction of new bone. Bioceramics mainly include biologically inert ceramics (such as alumina ceramics (Al_2O_3)), biodegradable ceramics (such as β-tricalcium phosphate (β-TCP)) and bioactive ceramics (such as hydroxyapatite (HAP), biological Active glass ceramics (BGC), and dual-phase calcium phosphorus ceramics (HA/TCP)).

3.4 Biomaterials and intermittent stem cells

3.4.1 HA/MSC composite

Hydroxyapatite is the most widely used calcium phosphate in dental and bone implant materials, mainly because it is similar to the mineral composition of natural bone and teeth, and has good biocompatibility and biological activity. In 1972, Japanese professor Hideki Aoki successfully synthesized hydroxyapatite and sintered it into ceramics. Subsequently, American professor Jarcho discovered that hydroxyapatite has good biocompatibility. Since then, scientists have carried out extensive basic research and clinical application research on hydroxyapatite, and it is recognized as a good bone repair material.

The molecular formula of Hydroxyapatite (HA) is $Ca_5(PO_4)_3(OH)$; however, in order to indicate that one HA unit cell unit consists of two molecules, the chemical formula is generally written as $Ca_{10}(PO_4)_6(OH)_2$. The HA crystal form belongs to the hexagonal crystal system, P63/m space group, which is characterized by six c-axes perpendicular to three equivalent a-axes (a1, a2, a3) at an angle of 120° to each other. Therefore, the crystallographic parameters of HA are $a = 9.418$ Å, $c = 6.881$ Å, and $\beta = 120°$. The unit cell of pure stoichiometric HA contains $10Ca^{2+}$, $6PO_4^{3-}$ and $2OH^-$. The PO_4^{3-} group exists as a tetrahedron with the P^{5+} ion as the center and 4 oxygen atoms as the top of the tetrahedron. From the (001) plane projection, 4 Ca^{2+} ions occupy Ca(I) sites: two at the z=0 level and two at the z=0.5 position, and then coordinate with the O of 9 PO43- groups to form a polyhedron. The other six Ca^{2+} ions have a coordination number of 7, which is expressed as Ca (II) sites: a group of three calcium atoms forms a triangle occupying z=0.25, and another group of three calcium atoms occupies z=0.75. 6 PO_4^{3-} groups surrounded by O and $1OH^-$. The two kinds of Ca(I) and Ca (II) have co-top or co-planar connections with the PO43-tetrahedron, so that the HA structure has better stability.

At the same time, hydroxyapatite can also change two different calcium sites by specific environment, so hydroxyapatite is also a highly nonstoichiometric calcium phosphate compound with a Ca/P molar ratio of 1.50–1.67. In addition, the hydroxyapatite crystal has two sets of crystal faces with different charging properties: the a and b faces are positively charged, and the c face is negatively charged. Due to the presence of positive Posner clusters in hydroxyapatite, the attraction of charges will make the c-plane the dominant growth plane, and finally, a hexagonal column structure of hydroxyapatite is obtained. Taking advantage of these properties, HA is also often used as an adsorption and separation material for proteins and heavy metals.

Synthetic hydroxyapatite ($Ca_{10}(PO_4)_6(OH)_2$, HA) is widely used in bone tissue engineering due to its similarity to the inorganic composition of bone, good biocompatibility, and osteoconductivity. HA prepared by traditional sintering lacks osteoinductive properties. Preliminary studies have shown that both the microstructure and nanostructure on the surface of hydroxyapatite ceramics can affect the adhesion, proliferation, and differentiation of osteoblasts, and implantation in vivo can promote osteogenesis. However, it is unclear how the micro- and nano-structures play a role in promoting osteogenesis, especially whether the micro- and nano-structures of hydroxyapatite bioceramics differ in their osteogenic effects. Whether the combination of micro- and nano-structures will produce more excellent osteogenic properties is the key to designing and optimizing bioactive bone repair materials. However, directly constructing patterned structures on the surface of hydroxyapatite ceramics is challenging due to the brittleness of ceramics. It is worth noting that, as a bone repair material used in clinical applications, it will be accompanied by corresponding degradation after implantation in the body, so whether the structure can produce sufficient active stimulation to cells to promote osteogenesis before the change of the structure due to degradation is also a bone graft. Factors such as this need to be considered in material design.

The modified HA nanoparticles prepared by Thian et al. [2] promoted the adhesion of osteoblasts and the expression of alkaline phosphatase, collagen type 1, and osteopontin by increasing the hydrophilicity of the surface. Shi et al. [3] found that both HA nanospheres and nanorods can promote cell proliferation and migration. Xia [4] also found that both micro- and nanostructures can promote the osteogenic differentiation of adipose stem cells. In addition, using the method of precursor hydrothermal conversion, Xia et al. [5] successfully prepared nano-structured HA porous scaffolds. The results showed that the nano-structure significantly promoted the adhesion, proliferation, and osteogenic differentiation of bone marrow mesenchymal stem cells. The material is applied in bone repair. Using wollastonite and α-TCP as hard templates, Xia et al. [6] used different hydrothermal media to obtain HA with different micro/nano-structures, and amorphous calcium phosphate can be rapidly transformed into micro/nano-structures in a simulated body fluid environment HA.

Interestingly, Lin [7] used α-TCP as a precursor for hydrothermal conversion to obtain HA with a micro/nano-composite structure. He found that the prepared nanosheets, nanorods, and a micro/nanorod-like combination of three structures can increase the selective adsorption of proteins on the surface of the material, and promote cell adhesion, proliferation, and differentiation; and compared with the two single-level nanostructures, the micro/nanorod-like composite structure has better performance. Yi et al. [8] combined the chelation of citrate and HA, the cross-linking of β-CD and citrate, and the in situ ROP of L-lactide. The resulting PLA-CD-HA-72 has a high grafting amount of PLA up to 11.7 wt%, and also exhibits several advantages such as increased particle dispersion, decreased hydrophilicity, and favorable positive zeta potential, relative to HA and β- CD/citrate modified HA. This endows PLA-CD-HA-72 with fairly good cell adhesion properties and osteoinductivity to MSCs. PLA-CD-HA-72 is expected to be a very promising biocompatible and bioactive material for bone fixation and/or bone tissue engineering.

Bian et al. [9] prepared collagen (Col)/nano-hydroxyapatite (nHA) composite nanofibers based on ordered nHA ceramic fibers by sintering electrospun PVP/HA composite fibers and then impregnating them with collagen solution. Compared with direct electrospinning, this method does not require strict control of temperature or humidity, prevents nHA from agglomerating during sintering, and reduces the use of expensive Col.

Composite nanofibrous membranes with bone-like components and structures have great potential for application in bone tissue engineering. And the method is easy to implement. The mineralization and degradation properties of nanofiber biocomposites are beneficial for osteoconduction and the growth and metabolism of MSCs. The protein adsorption of Col/HA nanofiber biocomposites is also excellent. There was no significant difference in cell viability between the experimental group and the control group, and the cytotoxicity of the Col/HA nanofiber biocomposite to MSCs was weak. Overall, this method successfully prepared Col/HA nanofiber biocomposites in a common electrospinning environment with uniformly distributed high content of nHA, low Col concentration, and less Col consumption, all of which make it an excellent choice for tissue engineering (Fig. 2).

Ressler et al. [10] used pH-responsive chitosan-hydroxyapatite hydrogel as a three-dimensional support for the osteogenic differentiation of encapsulated mesenchymal stem cells (MSCs). The linked hydrogel can serve as a three-dimensional support for the proliferation and differentiation of MSCs into osteoblasts. Expression of the characteristic bone genes Runx2, ALP, COLL I, OCN, and calcium phosphate deposits indicated that ECM

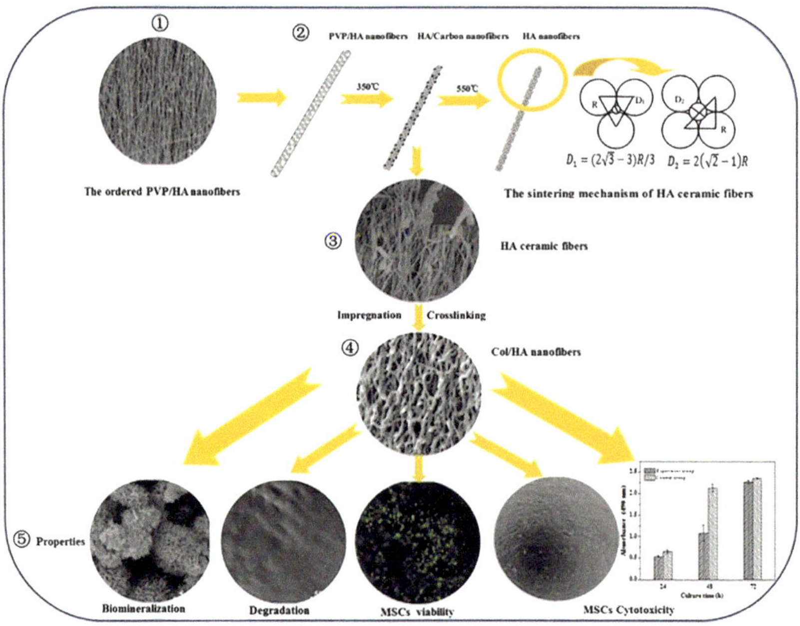

Fig. 2 Preparation and biological properties of collagen/nano-hydroxyapatite composite nanofibers based on ordered nano-hydroxyapatite ceramic fibers.

mineralization occurred during 14 days of cell culture. During cell culture, MSCs are uniformly dispersed throughout the Cht/HAp hydrogel, which is essential for complete bone healing. The prepared Cht/HAp hydrogels are highly stable under physiological and enzymatic conditions, supporting the osteogenesis of encapsulated mesenchymal stem cells. Kundu [11] investigated polycaprolactone (PCL) fibers combined with in situ mineralized HAP MMT-clay and MMT-clay using a pressurized gyratory (PG) device, using a new fabrication method to successfully fabricate HAP-nanoclay-PCL fibers.

Furthermore, 3D scaffolds made using the prepared fibers were able to enhance bone growth, cell viability, and proliferation. The results showed that the polymer fibrous scaffolds were biocompatible and cells were able to grow and differentiate on the fibrous scaffolds. Significant increases in cell viability, osteogenic differentiation, ECM formation, and collagen formation were observed using the PCL HAP MMT-clay fiber scaffold compared to the behavior in PCL fibers. Furthermore, intracellular ALP levels increased with PCL HAP MMT-clay fibrous scaffolds, indicating enhanced osteogenic differentiation of MSCs. This work shows the future promise of fabricating composite nanoclay polymer fibers, which are considered as scaffolds for bone tissue engineering applications.

3.4.2 Nano materials/MSC composite

Currently, the conventional treatments for malignant tumors include chemotherapy and surgical resection. However, conventional chemotherapeutic approaches often lack selectivity and have off-target toxicities, as only a small percentage of antitumor drugs end up in tumor lesions, resulting in the same damage being caused to healthy tissue by chemotherapy. In addition, surgical resection cannot eradicate diffuse tumor microresidues, and the prognosis is poor, as this may easily lead to secondary recurrence [12]. Therefore, the development of effective chemotherapeutic drug delivery therapies to treat malignant tumors and their timely monitoring has attracted great interest among scientists.

The most difficult problem with drug delivery systems is the inability of the drug to be efficiently delivered to the intended biological target. Therefore, the development of delivery vehicles that can effectively target and reach tumor lesions and stably release various anticancer drugs is the ultimate goal of designing novel drug delivery systems [13,14]. Current research shows that tumor-oriented stem cells, especially mesenchymal stem cells

(MSCs), can be used as drug carriers to provide a good drug delivery system for tumor therapy, thereby improving the efficiency of tumor-targeted therapy and reducing the dosages of chemotherapeutic drugs. Systemic toxicity stem cells have been reported to have a strong intrinsic tumor tropism in response to tumor-secreted chemoattractants, angiogenic factors, and inflammatory factors, making them effective candidates for targeted drug delivery vehicles [15]. In addition, stem cells have been shown to be loaded as nanocarriers, enabling efficient drug delivery throughout the tumor and exerting strong antitumor efficacy [16].

It has been reported that MSCs can carry RGD-modified doxorubicin (DOX)-polymer conjugate (PPCD) to target glioma, which has a good antitumor effect on metastatic specific glioma [13]. In addition, several preclinical studies have demonstrated that stem cell vectors can be administered by intratumoral, peritumoral, intravenous, and intraarterial injections. Human adipose stem cells (hADSCs) are a new type of MSC. hADSCs can be easily isolated from adipose tissue, cultured and expanded in vitro, and can be used as a promising delivery vehicle [17]. Therefore, this makes it possible to autograft hADSCs into patients while avoiding immune rejection.

Designing excellent nanocarriers for antitumor drugs is very important for drug delivery systems. It has been reported that drug-loaded engineered nanoparticles have demonstrated effective drug-loading capacity and controlled release performance, which can effectively improve drug utilization while reducing unnecessary drug consumption [18,19]. Inorganic nanoparticles, such as iron oxide [20], gold [21], and silica [22], can be engineered into different sizes, structures, and geometries, which endow them with unique physical, electrical, magnetic, and optical properties. As inorganic nanoparticles have been designed in more and more specific ways, they have been widely used in drug delivery and molecular imaging applications.

Mesoporous silica nanoparticles (MSN) with large specific surface area have good adsorption capacity and biocompatibility, and are certified by the United States Food and Drug Administration (FDA) as a safe carrier for drug delivery. Wagner et al. used MSN as a pH-responsive drug carrier for the anticancer immunostimulator R848, which could ensure controlled drug release. Another common inorganic nanoparticle that has been successfully used as a contrast agent and delivery vehicle is the superparamagnetic Fe_3O_4 nanoparticle, which can be engineered to a specific size. As a kind of superparamagnetic nanoparticle (MNPs), Fe_3O_4-based nanocarriers can achieve enhanced imaging of tumors by magnetic resonance imaging (MRI), which is a noninvasive detection method and is widely used in clinical translation MRI technology MNPs-based carriers can be localized to achieve simultaneous detection in the early stage of drug delivery and late stage of metabolism, so as to help determine the therapeutic effect and potential toxicity of the carrier in vivo. Therefore, the combination of nanomaterials MSNs and Fe_3O_4 nanoparticles can creatively prepare pH-sensitive and sustainable nanocarriers based on small-molecule drugs with excellent MRI imaging ability, which can monitor the drug delivery process in time.

There are many factors that direct the regulation of stem cell differentiation, such as bone morphogenetic protein-2 (BMP-2), BMP-7, miRNA-346, dexamethasone (DEX), etc. [23,24]. However, the instability and easy degradation of these substances limit the application of these factors. Advances in nanotechnology have led to the development of advanced biomaterials with custom combinations of properties. A series of biomedical nanomaterials have been developed to mimic tissue complexity and modulate stem cell fate [25]. Studies have shown that the chemical composition, surface morphology, mechanical properties, electrical properties, and morphological properties of nanomaterials significantly affect the fate of stem cells [26].

Nanomaterials come in many forms, such as nanofibers, nanopillar arrays, and nanoparticles. The interaction of nanoparticles (NPs) with stem cells plays an important role in diagnostic, drug delivery, and therapeutic applications. NPs can easily translocate across cell membranes and distribute in the cytoplasm or lysosomes, thereby activating or inhibiting certain cellular signaling pathways and inducing or inhibiting differentiation. The physicochemical properties, such as size, shape, and surface chemistry, of AuNPs may affect the multidirectional differentiation of stem cells [27]. BSA-coated gold nanospheres (ball-40, ball-70, ball-110), gold nanostars (star-40, star-70, star-110), and gold nanorods (rod-40, rod-70, rod-110) were prepared. Subsequently, their effect on osteogenic differentiation of mesenchymal stem cells was observed [28]. The study found that ball-40, ball-70, and rod-70 significantly promoted osteogenic differentiation, while rod-40 significantly inhibited osteogenic differentiation. Star-40, star-70, star-110, ball-110, and rod-110 had negligible effects on osteogenic differentiation [28]. Furthermore, AuNPs were constructed with functional groups of amine (AuNP-NH2, 21.76 ± 3.93 nm), carboxyl group (AuNP-COOH, 17.08 ± 2.28 nm), and hydroxyl group (AuNP-OH, 12.42 ± 1.33 nm). AuNP-NH_2 and AuNP-OH did not inhibit osteogenic differentiation, whereas AuNP-COOH significantly reduced ALP activity and matrix mineralization in MSCs by upregulating the expression of TGF-b and FGF-2, thereby enhancing osteogenic differentiated cells proliferation [29].

Zinc oxide NPs are another metal oxide nanoparticle that has a wide range of applications. In recent years, the interaction between ZnO NPs and stem cells has been investigated. Zinc is an essential trace element for the human body. Zinc deficiency reduces bone mass, delays bone metabolism, and results in further retardation of bone growth. Zinc promotes bone formation and mineralization both in vitro and in vivo [30]. Studies have shown that Zn can promote the proliferation and differentiation of osteogenic MC3T3-E1 cells, as well as protein synthesis. Furthermore, the effect of ZnO NPs on osteogenic differentiation is dose-dependent. 30 μg mL^{-1} ZnO NPs promoted the osteogenic differentiation of mesenchymal stem cells, while 60 μg mL-1 ZnO NPs had no significant effect on the osteogenic differentiation of mesenchymal stem cells [31]. Therefore, nanoparticles can not only induce stem cell differentiation by themselves, but can also be modified by different chemical bonds to regulate stem cell fate.

In addition, nanoparticles can be used to load drugs or other factors to achieve sustained release of drugs, thereby regulating stem cell differentiation. For example, BMP-2 peptide was functionalized and covalently grafted on the surface of mesoporous silica nanoparticles (MSNs-pep) via aminosilane linkers, and then DEX was loaded into silica nanoparticles to construct a nano-osteogenic delivery system (DEX@MSNs-pep) [32]. The results of in vitro experiments showed that MSNs-pep promoted the osteogenic differentiation of BMSCs in terms of alkaline phosphatase activity level, calcium deposition, and bone-related protein expression. In addition, the addition of DEX to MSNs-pep could further promote the osteogenic differentiation of MSCs. After 3 weeks of intramuscular implantation in rats, computed tomography (CT) images and histological examinations demonstrated that the nano-osteogenic delivery system induced efficient osteoblast differentiation and bone regeneration in vivo.

Stem cells can differentiate into other types of cells, so stem cells have been widely used in the medical field, especially in regenerative medicine [33]. Stem cells have some special properties, such as tumorigenicity, so they can be used in the field of drug delivery. Likewise, nanoparticles that camouflage stem cell membranes showed similar targeting functions to tumors. In recent years, stem cell membrane encapsulation has been used as a strategy to improve light-based applications [34]. Stem cell membranes are derived from bone marrow-derived mesenchymal stem cell membranes.

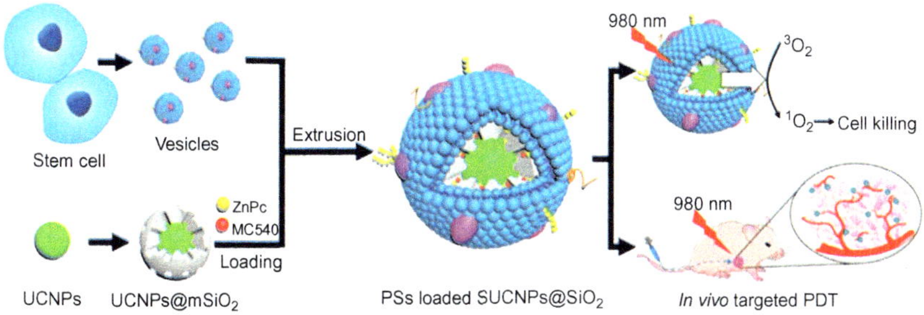

Fig. 3 Fabrication process of SUCNPs@mSiO_2 and its mechanism in PDT.

Upconversion nanoparticles (UCNPs) were first coated on a mesoporous silica shell, then two photosensitizers (PSs) were loaded onto the shell, and the final nanoparticles were obtained by stem cell membrane coating. Nanoparticles disguised as stem cell membranes exhibited good cytocompatibility and hemocompatibility when cultured with normal hepatocytes or mouse blood (Fig. 3). Furthermore, compared with bare nanoparticles, the uptake of nanoparticles disguised as stem cell membranes was enhanced when incubated with HeLa cells, suggesting the ability of stem cell membranes to target cancer cells. When injected into tumor-bearing mice, the stem cell membrane-encapsulated nanoparticles accumulated significantly more in tumors than bare nanoparticles. At the same time, there was less accumulation in the liver, indicating reduced clearance of nanoparticles that camouflage stem cell membranes by MPS.

Nanoparticle injection combined with laser irradiation can significantly improve the photodynamic antitumor efficacy, completely control tumor growth, and have no obvious side effects. Furthermore, in addition to the characteristics of targeting tumors, stem cell membranes also have the function of accumulating to the same type of cells; that is, targeting homologous cells. Therefore, nanoparticles can be encapsulated with the same cell membrane as the target cells, allowing the nanoparticles to have immune escape, long circulation time, and targeting functions.

The interaction between stem cells and nanomaterials is very complex. Cells can reshape their microenvironment to a certain extent, and materials with micro/nano-structures can pass their own properties, such as adhesion, hardness, micro- and nano-structures, etc., to regulate the behavior of thousands of cells. Stem cells cultured on materials can sense the properties of materials, integrate effective growth information through signal transmission, and finally, apply the integrated information to the regulation of cell fate. Currently, it remains challenging to fully discover and understand the mechanisms by which stem cells respond to specific material properties due to their environment with highly complex multicomponent signaling. But there is no doubt that it has been confirmed by a large number of studies that micro- and nano-structured materials with specific properties play a decisive role in the adhesion, spreading, proliferation, and differentiation of stem cells.

Further understanding the development of mechanotransduction between stem cells and nanomaterials will have a major impact on the fields of tissue engineering and regenerative medicine. With the development of micro/nano synthesis and processing technologies such as nano-assembly, nano-etching, and nano-imprinting technology, the preparation methods of ordered micro/nano structures have gradually matured. Therefore, people began to pay attention to the related research to explore the influence of regular changes of micro/nano-structure on stem subdivision.

Numerous studies have confirmed the interaction between stem cells and nanoscale geometric features [35]. Dalby et al. explored the effect of the order of nanopore arrays on the differentiation of MSCs [36]. They prepared a series of nanopore arrays with different degrees of order based on polymethyl methacrylate (PMMA) and found that the highly ordered nanopore structures adhered to cells without using any growth factors. The effect on osteogenic differentiation was not obvious, while the cells on the disordered random nanopore structure showed more osteoblast morphology after 14 days of culture, and the content of osteogenic marker proteins also increased, indicating that, based on the geometric morphological characteristics, change-induced stem cell differentiation has good research value and application prospects.

Kim et al. studied the effect of nanopore diameter on polystyrene substrates on pancreatic differentiation of human embryonic stem cells, and confirmed that polystyrene nanopatterned surfaces can significantly promote pancreatic differentiation of human embryonic and induced pluripotent stem cells [37]. The comparison found that the 200 nm diameter nanopores highly upregulated the expression of PDX1, a key transcription factor for pancreatic development, and the differentiation rate of pancreatic progenitor cells was approximately three-times higher than that of the flat control group. Furthermore, the 200 nm nano-diameter nanopore substrate significantly enhanced the efficiency of pancreatic endocrine cells to produce insulin, glucagon, or somatostatin in the presence of biochemical factors. The study suggests that combining cytokines with nanogeometric topography may accelerate the development and application of cell therapy.

Nanopillar arrays are also common micro/nano-structures used in tissue engineering. Many studies have prepared nanopillar arrays and used this structure as a substrate to regulate the differentiation of stem cells. Qm et al. reported the regulatory effect of TiO_2 nanopillar arrays on the osteogenic differentiation of MSCs [38]. They found that the array significantly promoted the osteogenic differentiation of MSCs, confirming the regulatory effect of the nanoarray on stem cell differentiation. At the same time, the authors further confirmed that the osteogenic differentiation of MSCs in the TiO_2 nanoarray region was significantly enhanced by micro-patterning of TiO_2 nanoarrays, while the stem cell phenotype was maintained in the TiO_2 planar region, confirming the huge effect of nanoarrays in site-directed differentiation.

Liu et al. prepared a series of polyglycolide micropillar arrays with different aspect ratios and found that the osteogenic differentiation of bone marrow mesenchymal stem cells was enhanced with increasing aspect ratios after induction of osteogenic or adipogenic differentiation for several days, while adipogenesis decreased with increasing aspect ratio [39]. In addition, the shape of the nucleus also changes with the change of the aspect ratio of the micropillars, and the deformation becomes more significant with the increase of the length of the micropillars. Therefore, it is proposed that the geometrical deformation of the nucleus may also be an important factor affecting the differentiation of stem cells.

The integration of nanobiomaterials with stem cells represents a promising strategy for the treatment of myocardial infarction. While stem cells and nanobiomaterials have each shown partial success in cardiac repair, the therapeutic efficacy of each has been low in clinical settings. Therefore, the combination of nanobiomaterials and stem cells is being actively researched to produce a synergistic effect in the treatment of myocardial infarction (MI). To date, various types of nanomaterials have been combined with stem cells to control cell fate, modulate the therapeutic behavior of stem cells, and make them more suitable for cardiac repair. Although stem cell therapy has emerged as a promising treatment option for MI, its therapeutic benefit remains limited due to poor engraftment and survival of implanted cells. Efficient delivery and prolonged survival of stem cells present significant challenges for their application in the treatment of MI. However, biomimetic self-assembled peptide nanofibers can be used as cell delivery vehicles for the treatment of myocardial infarction. Due to their biocompatibility and biodegradability, self-assembling peptides are used not only for drug delivery but also for cellular delivery of cardiac repair. Previous studies utilizing self-assembling peptides as cell delivery vehicles demonstrated greatly improved survival of engrafted cells and enhanced myocardial regeneration after MI. This is mainly because self-assembling peptides can increase cell retention and promote cell survival after in vivo delivery.

3.4.3 Biomaterials and derivatives of MSCs

Mesenchymal stem/stromal cells (MSCs) have been identified as key players in bone homeostasis and are considered a promising therapy for diseases such as osteoarthritis (OA) and rheumatoid arthritis (RA). However, some known drawbacks limit their use, especially ethical concerns and immune rejection. Therefore, a byproduct of MSCs, extracellular vesicles (EVs), is emerging as a potential solution to overcome some of the problems of protocells. EVs can be modulated by cell pretreatment or vesicle engineering, and thus, represent a plastic tool that can be used in regenerative medicine (Fig. 4). Furthermore, the use of biomaterials is important for improving EV delivery and indirectly regulating their content and secretion. Studies focusing on intercellular communication have used body fluids such as saliva, urine, blood, and breast milk to show that all cell types can release lipid bilayer vesicles, which are now commonly labeled as EVs [40,41]. It is well known that EVs naturally carry proteins, lipids, and nucleic acids and also have potential for bioengineering as drug delivery systems [42,43]. Depending on the EV cargo and target cell type, EVs can affect various cellular processes, more specifically, proliferation, differentiation, senescence, and apoptosis, while also affecting immune regulation, blood coagulation, and angiogenesis [44,45].

Currently, the International Society for Extracellular Vesicles (ISEV) proposes to distinguish three main types of EVs based on size and release mechanism: apoptotic bodies, microvesicles, and exosomes [41]. Apoptotic bodies are the largest, ranging from 50 m to 5 μm. In the case of programmed cell death, they are formed by blebbing of the plasma membrane [46]. Microvesicles (MVs) are sprouted from the plasma membrane and their size is between 50 nm and 1 μm. Finally, exosomes are typically less than 150 nm in diameter and result from the fusion of multivesicular bodies (MVBs) with the plasma membrane, releasing intraluminal vesicles (ILVs) into the surrounding microenvironment, hence the term exosomes [41].

On their surfaces, EVs can also display several molecules that directly affect target cells, without the need for uptake. Immunoglobulins, complement proteins, coagulation factors, cytokines, enzymes, and DNA associated with vesicle membranes have been detected [47]. In the context of RA,

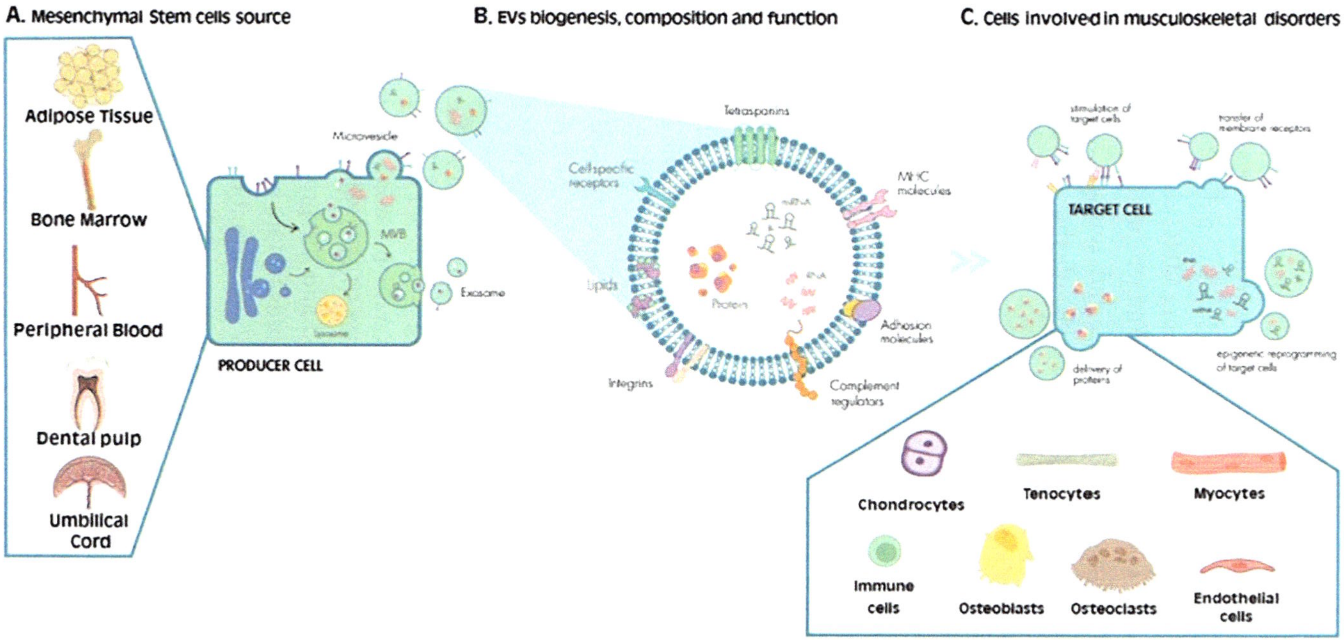

Fig. 4 Cell-to-cell Communication via EVs in the musculoskeletal system.

Cloutier et al. demonstrated that synovial fluid contains platelet-derived EVs, showing immunoglobulins, antigens, and complement proteins as immune complexes capable of exerting proinflammatory signals in the presence of neutrophils [48]. Finally, EVs can also serve as antigen-presenting vehicles, as shown by Raposo et al., who reported that murine and human B lymphocytes release vesicles containing MHC class II molecules that can activate specific T-cell responses [49]. Antigen-presenting EVs can also stimulate T-cell and macrophage phenotype shifts towards Treg and M2, respectively, thereby shortening the length of the inflammatory phase in musculoskeletal regeneration [46].

Immune cells have been shown to transfer membrane receptors to other cells. For example, leukocyte-derived EVs can transfer monocyte/macrophage tissue factor to platelets. Conversely, platelet-specific adhesion molecules can be relocated to hematopoietic cells through the action of EVs. Additional functions of EVs include epigenetic reprogramming of target cells by altering DNA methylation patterns, histone modifications, and posttranscriptional editing of noncoding RNAs. As mentioned above, with regard to MSCs, the delivery of EVs is also an important factor to consider. Administration of so-called "naked EVs," i.e., without any coating, results in nonspecific binding to nontarget organs, which in turn leads to side effects and reduced concentrations at the site of injury, thereby affecting therapeutic efficacy [50]. Therefore, combining with biomaterials is considered as a feasible strategy to ensure proper targeting, preservation, and controlled release of therapeutics [51].

The reported results of using encapsulated MSC-EVs in hydrogels to promote regeneration are encouraging. Mardpour et al. achieved sustained EV release and resistance to degradation by using polyethylene glycol (PEG) hydrogels in a rat liver regeneration model. This study showed the feasibility of the delivery method and the pro-regenerative effect of MSC-EVs on damaged liver [52]. A study by Qin et al. chose a commercial hydrogel-HyStem-HP-to deliver EVs to promote osteogenesis in a rat model of critical-size bone defect and reported that this system was responsible for promoting bone formation [53]. By combining EVs with 3D devices made of cartilaginous ECM and gelatin methacrylate hydrogels [54], and vesicles into photo-induced imine cross-linking (PIC) hydrogel glues [55], the technique has also been shown to be effective in cartilage regeneration.

3.5 Other aspects

3.5.1 MSCs detect nanomaterials

In the field of stem cell therapy, mesenchymal stem cells (MSCs) have shown great potential in the treatment of various diseases, such as heart disease [56–58], cerebral ischemia [59], and acute liver failure, and in bone regeneration [60,61]. In vivo noninvasive and highly sensitive tracking of MSCs in the host is a key problem that still needs to be solved in stem cell therapy. Tracing of transplanted MSCs can help to determine the location and number of MSCs, and to monitor the long-term fate and distribution of MSCs after transplantation [62]. Therefore, high-resolution imaging techniques are needed to monitor the location of transplanted MSCs in mice in vivo to better understand the behavior of MSCs after transplantation. Currently commonly used cell tracking methods include internal labeling methods such as reporter gene-modified green fluorescent protein (GFP) or luciferase [63,64]. There are also some external labeling modalities, such as quantum dot (QD)-based fluorescence imaging [65,66], magnetic resonance imaging (MRI) [67,68], computed tomography (CT) imaging [69], and photoacoustic (PA) imaging [70,71], and other in vivo tracking modalities. However, due to the inherent limitations of existing imaging methods and tracer contrast agents, these imaging methods cannot meet the demand for high sensitivity and high spatiotemporal resolution for in vivo imaging.

In recent years, surface-enhanced Raman (SERS) spectroscopy has attracted extensive attention as a noninvasive and highly sensitive imaging technique due to its narrow peaks and rich fingerprint information. Compared with spectral techniques such as infrared and fluorescence, SERS also has the advantages of narrow spectral lines, less quenching, less damage to biological samples, and no interference from background fluorescence. It is especially suitable for the study of cells or tissues. The differentiation process of stem cells is irreversible and involves many biochemical changes. Real-time and efficient detection of these biochemical changes by SERS has unique advantages. Notingher et al. [72,73] used SERS to study the in vitro differentiation process of mouse embryonic stem cells. They found that changes in intracellular material after differentiation manifested as a decrease in RNA and DNA content. Chan et al. [74] also studied the process of human embryonic stem cells differentiation into muscle cells. With the help of multivariate statistical analysis tools, they achieved precise Raman spectroscopy identification of stem cell-differentiated cardiomyocytes and fetal left atrial cardiomyocytes. In addition, SERS studies on the differentiation process of bone marrow mesenchymal stem cells from humans, mice, and monkeys have also reported that, during the differentiation process of stem cells, the spectral changes are generally manifested as a decrease in the intensity of RNA and DNA-related vibrational peaks. This is related to the storage of a large amount of nucleic acid substances before stem cell differentiation.

There are also some special differentiation processes, such as the differentiation of bone marrow mesenchymal stem cells into osteoblasts, and the Raman spectrum shows an increase in the intensity of phosphate-related vibrational peaks (950–970 cm^{-1}). Hunfner et al. [75] prepared a nucleus-targeted SERS probe to identify differentiated and undifferentiated neuronal stem cells. They found that DNA/RNA and transcriptional proteins increased significantly during stem cell differentiation. Cao et al. [76] constructed a label-free SERS probe to detect the differentiation of lung stem cells (LR-MSCs). The label-free SERS probe consists of three parts: gold nanostars as a reinforcing substrate, thiolated polyethylene glycol (SH-PEG-COOH) as a protective agent and stabilizer, and a cell membrane penetrating peptide (TAT) molecule, as membrane penetrating ribs were coupled to the surface of gold nanostars to increase the transmembrane transport capacity of nanoparticles. The stability, biocompatibility, and interaction between SERS probes and cells were evaluated. *Trans*-Well technology was used to construct a co-culture induced differentiation system of LR-MSCs with alveolar n-type epithelial cells or transforming growth factor (TGF-P). The differentiation of LR-MSCs was detected and differentiated by SERS. The results show that TAT-conjugated gold nanostars can be used as a label-free SERS probe for research in the field of stem cell differentiation.

Evidence suggests that EVs specifically released from mesenchymal stromal cells (MSCs) are potent cell-free regeneration agents. However, for MSC EVs to be used in therapeutic practice, there must be a standardized and reproducible method to characterize them. The detection and characterization of EVs is a challenge due to their nanoscale size and molecular heterogeneity. To address this challenge, gold nanohole arrays of different sizes and shapes were fabricated by electron beam lithography. These platforms serve the dual purpose of capturing individual EVs and enhancing their vibrational signatures in surface-enhanced Raman spectroscopy (SERS).

Nina [77] reported SERS spectra of MSC EVs derived from pancreatic tissue (Panc-MSC) and bone marrow (BM-MSC). Principal component analysis (PCA) was used to determine the principal component differences between these two groups at 1236, 761, and 1528 cm^{-1}, corresponding to amide III, tryptophan and in-plane —C=C— vibrations, respectively. Furthermore, several machine learning methods were used to differentiate BM and PancMSC EVs and achieved 89% accuracy, 89% sensitivity, and 88% specificity using logistic regression. Adrianna et al. [78] employed noninvasive surface-enhanced Raman scattering (SERS) to study interstitial stem cells to better understand cellular processes in situ. Long-term differentiation of MSCs on biocompatible gold nanoisland SERS substrates was demonstrated, combining cellular imaging with spectroscopic detection of molecular species and chemical events that occur on cell membranes near the surface of SERS substrates. Multiple signs of bone tissue formation were detected, from early stages to mature osteoblasts, without labeling, demonstrating that SERS imaging alone is sufficient to study the osteogenic differentiation process of such cells, paving the way for serial label-free screening of living cells the road.

Jaehan and collaborators [79] stacked three-dimensional cell culture scaffolds and surface-enhanced Raman spectroscopy (SERS)-based biosensors to form a multilayered system, which allows monitoring of protein markers secreted by cultured stem cells without the need for periodic cell and/or or media collection. The cell culture scaffold supports the proliferation and osteogenic differentiation of adipose-derived mesenchymal stem cells (ADSCs). The protein labels detected by the SERS capture substrate were combined with SERS tags made of Au-Ag alloy nanoboxes. Incorporating different Raman reporter genes into SERS tags allows easy identification of target proteins for multiplexing. The resulting SERS-based immunoassay can detect protein markers at pg/mL levels without crosstalk and interference (Fig. 5).

Using MSCs for the repair of tissue damage, SERS imaging enables monitoring of the treatment process of MSCs, bringing new opportunities and hope for stem cell tracking and treatment research.

3.5.2 MSCs oxidative stress regulator nanomaterials-nanozymes

Mesenchymal stem cell (MSC)-based therapies have great potential to treat a variety of diseases because of their ability to heal injured tissues/organs through paracrine factor secretion and immune modulation. Despite high hopes, the low survival of transplanted cells in damaged tissues due to

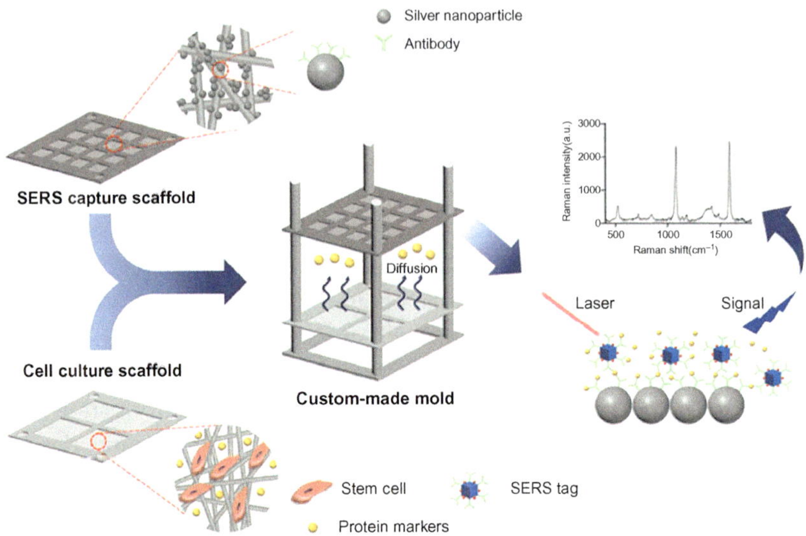

Fig. 5 Schematic illustration of integrating 3D cell culture scaffold and SERS-based biosensor for the detection of protein markers secreted from stem cells.

elevated levels of oxidative stress remains the greatest obstacle to MSC-based cell therapy.

To achieve the desired therapeutic efficiency, it is necessary to ensure the survival of transplanted MSCs in an environment of high oxidative stress. Therefore, combining ROS-scavenging nanomaterials with MSCs can improve the therapeutic efficiency of cell therapy. Sahu et al. [80] proposed the use of ROS-scavenging nanozymes to protect transplanted MSCs from oxidative stress-mediated apoptosis, thereby improving therapeutic efficacy. Prussian blue (PB) nanoparticles were incorporated into MSCs as a biocompatible ROS-scavenging nanozyme without affecting the stringency and differentiation potential of MSCs. They synthesized biocompatible PB nanozymes that could be efficiently incorporated into MSCs without affecting the biological activity of stem cells. Incorporation of PB nanozymes not only enhanced cell survival under conditions of high oxidative stress, but also enhanced cellular paracrine and antiinflammatory properties. Administration of PB-MSCs to an animal model of hepatic I/R injury resulted in significantly less tissue damage compared to normal MSCs, resulting in profound in vivo therapeutic effects in an animal model of hepatic ischemia-reperfusion (I/R) injury. The findings suggest that the incorporation of nanozymes into MSCs is a simple and effective approach to enhance the therapeutic potential of MSC-based cell therapy (Fig. 6A).

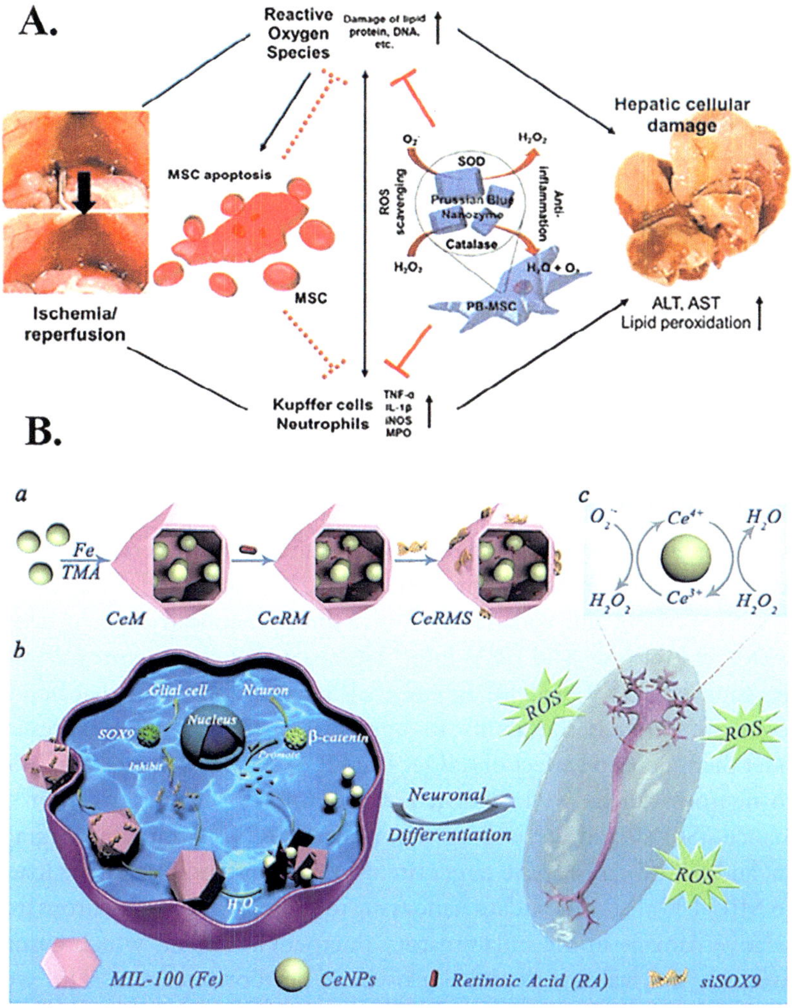

Fig. 6 (A) Schematic illustration of neuron directional differentiation and neuroprotection for enhanced neurogenesis. (B) Schematic illustration of neuron directional differentiation and neuroprotection for enhanced neurogenesis.

It has been reported that exogenous neural stem cell (NSC) transplantation contributes to self-repair after neuronal loss. Current approaches to NSC transplantation mainly focus on neuronal directed differentiation by developing new physical, chemical, and biological methods to promote neural differentiation. However, brain cells of AD patients always suffer from pathologically high levels of oxidative stress. Sustained oxidative stress may lead to exogenous NSC death and a cascade of neuronal apoptosis, restricted synaptic growth, gliosis, cognitive impairment, and ultimately, behavioral dysfunction. Several studies report using optimized NSC microenvironment components to enhance NSC viability and engraftment effect. However, due to persistent oxidative stress, an unfavorable inflammatory microenvironment would be detrimental to the survival of exogenous NSCs and newborn neurons, impairing therapeutic efficacy. Therefore, it is necessary to eradicate the over-produced ROS in the inflammatory microenvironment to promote nerve regeneration.

Cerium dioxide was chosen as a typical and credible antioxidant nanozyme because of its admirable ROS scavenging activity and excellent biocompatibility. Based on the electron transfer between cerium (III) and cerium (IV), their excellent superoxide dismutase (SOD) and catalase (CAT) mimetic activities are expected to eliminate ROS in NSCs and newborn neurons [81]. Given its ability to reduce oxidative stress, ceria is able to provide long-term protection during NSC engraftment, differentiation, and nerve regeneration. Based on this, Qu et al. [82] designed ceria-doped MIL-100 with excellent antioxidant activity as a unique nanocarrier to co-deliver siSOX9 and retinoic acid (RA) into NSCs. siSOX9 can downregulate the expression of SOX9 protein, an essential transcription factor that benefits gliogenesis. RA has the ability to upregulate neuronal gene expression. Under the synergistic effect of siSOX9 and RA, the differentiation of NSCs into neurons can be well regulated. Furthermore, MOF as a carrier can enhance siSOX9 and RA cell uptake, protect siRNA from rapid degradation, and release cargo at the target site. More importantly, ceria was loaded into MIL-100 as an antioxidant nanozyme to protect newborn neurons from oxidative damage in the inflammatory microenvironment, which ensured satisfactory survival and neuronal growth of newborn neurons. This well-designed nanoplatform can be applied as a promising tool for directed neuronal differentiation and neuroprotection (Fig. 6B).

Acknowledgments

This work was supported by grants from the National Natural Science Foundation of China (11635013, 82260031), Research Fund for International Young Scientist (RFIS-1, E22ADI35501), supported by Jiangxi Provincial Natural Science Foundation (20224BAB206077, 20212BAB216073), the Project Youth Fund, supported by Shandong Provincial Natural Science Foundation (ZR2020QC097), the Nonprofit Central Research Institute Fund of Chinese Academy of Medical Sciences (2019PT320005), and The 2021 Central-Guided Local Science and Technology Development Fund (ZYYDDFFZZJ-1),

Gansu Provincial Hospital Intra-Hospital Research Fund Project (22GSSYB-6, 21GSSYB-8, 20GSSY5-2), The 2022 Master/Doctor/Postdoctoral program of NHC Key Laboratory of Diagnosis and Therapy of Gastrointestinal Tumor (NHCDP2022004, NHCDP2022008, NHCDP2022014), Key project funded by Department of Science and Technology of Shangrao City (2022A001), Double Thousand Plan of Jiangxi Province (2022 to LSZ).

References

[1] Wang DL. The birth and development of tissue engineering. China Biotechnol 2014;34(5):122–9.

[2] Thian ES, Ahmad Z, Huang J, Edirisinghe MJ, Jayasinghe SN, Ireland DC, Brooks RA, Rushton N, Bonfield W, Best SM. The role of surface wettability and surface charge of electrosprayed nanoapatites on the behaviour of osteoblasts. Acta Biomater 2010;6(3):750–5.

[3] Shi F, Liu Y, Zhi W, Xiao D, Li H, Duan K, Qu S, Weng J. The synergistic effect of micro/nano-structured and Cu(2+)-doped hydroxyapatite particles to promote osteoblast viability and antibacterial activity. Biomed Mater 2017;12(3), 035006.

[4] Xia L, Lin K, Jiang X, Fang B, Xu Y, Liu J, Zeng D, Zhang M, Zhang X, Chang J, Zhang Z. Effect of nano-structured bioceramic surface on osteogenic differentiation of adipose derived stem cells. Biomaterials 2014;35(30):8514–27.

[5] Xia L, Lin K, Jiang X, Xu Y, Zhang M, Chang J, Zhang Z. Enhanced osteogenesis through nano-structured surface design of macroporous hydroxyapatite bioceramic scaffolds via activation of ERK and p38 MAPK signaling pathways. J Mater Chem B 2013;1(40):5403–16.

[6] Xia L, Zhang N, Wang X, Zhou Y, Mao L, Liu J, Jiang X, Zhang Z, Chang J, Lin K, Fang B. The synergetic effect of nano-structures and silicon-substitution on the properties of hydroxyapatite scaffolds for bone regeneration. J Mater Chem B 2016;4(19):3313–23.

[7] Lin K, Xia L, Gan J, Zhang Z, Chen H, Jiang X, Chang J. Tailoring the nanostructured surfaces of hydroxyapatite bioceramics to promote protein adsorption, osteoblast growth, and osteogenic differentiation. ACS Appl Mater Interfaces 2013;5(16):8008–17.

[8] Yi W-J, Li L-J, He H, Hao Z, Liu B, Chao Z-S, Shen Y. Synthesis of poly(l-lactide)/β-cyclodextrin/citrate network modified hydroxyapatite and its biomedical properties. New J Chem 2018;42(18):14729–32.

[9] Bian T, Zhang H, Xing H. Preparation and biological properties of collagen/nano-hydroxyapatite composite nanofibers based on ordered nano-hydroxyapatite ceramic fibers. Colloids Surf A Physicochem Eng Asp 2020;602, 124802.

[10] Ressler A, Rodenas-Rochina J, Ivankovic M, Ivankovic H, Rogina A, Gallego Ferrer G. Injectable chitosan-hydroxyapatite hydrogels promote the osteogenic differentiation of mesenchymal stem cells. Carbohydr Polym 2018;197:469–77.

[11] Kundu K, Afshar A, Katti DR, Edirisinghe M, Katti KS. Composite nanoclay-hydroxyapatite-polymer fiber scaffolds for bone tissue engineering manufactured using pressurized gyration. Compos Sci Technol 2021;202, 108598.

[12] Poon W, Kingston BR, Ouyang B, Ngo W, Chan WCW. A framework for designing delivery systems. Nat Nanotechnol 2020;15(10):819–29.

[13] Zhang X, Yao S, Liu C, Jiang Y. Tumor tropic delivery of doxorubicin-polymer conjugates using mesenchymal stem cells for glioma therapy. Biomaterials 2015;39:269–81.

[14] Wu M, Zhang H, Tie C, Yan C, Deng Z, Wan Q, Liu X, Yan F, Zheng H. MR imaging tracking of inflammation-activatable engineered neutrophils for targeted therapy of surgically treated glioma. Nat Commun 2018;9(1):4777.

[15] Huang X, Zhang F, Wang Y. Design considerations of iron-based nanoclusters for non-invasive tracking of mesenchymal stem cell homing. ACS Nano 2014;8(5):4403–14.
[16] Zhang B, Yan W, Zhu Y, Yang W, Le W, Chen B, Zhu R, Cheng L. Nanomaterials in neural-stem-cell-mediated regenerative medicine: imaging and treatment of neurological diseases. Adv Mater 2018;30(17), e1705694.
[17] Kim T, Momin E, Choi J, Yuan K, Zaidi H, Kim J, Park M, Lee N, McMahon MT, Quinones-Hinojosa A, Bulte JW, Hyeon T, Gilad AA. Mesoporous silica-coated hollow manganese oxide nanoparticles as positive T1 contrast agents for labeling and MRI tracking of adipose-derived mesenchymal stem cells. J Am Chem Soc 2011;133 (9):2955–61.
[18] Huang P, Bao L, Zhang C, Lin J, Luo T, Yang D, He M, Li Z, Gao G, Gao B, Fu S, Cui D. Folic acid-conjugated silica-modified gold nanorods for X-ray/CT imaging-guided dual-mode radiation and photo-thermal therapy. Biomaterials 2011;32(36):9796–809.
[19] Zhang W, Wang M, Tang W, Wen R, Zhou S, Lee C, Wang H, Jiang W, Delahunty IM, Zhen Z, Chen H, Chapman M, Wu Z, Howerth EW, Cai H, Li Z, Xie J. - Nanoparticle-laden macrophages for tumor-tropic drug delivery. Adv Mater 2018;30 (50), e1805557.
[20] Jun YW, Lee JH, Cheon J. Chemical design of nanoparticle probes for high-performance magnetic resonance imaging. Angew Chem Int Ed Eng 2008;47 (28):5122–35.
[21] Kumar A, Huo S, Zhang X. Neuropilin-1-targeted gold nanoparticles enhance therapeutic efficacy of platinum (IV) drug for prostate cancer treatment. ACS Nano 2014;8 (5):4205–20.
[22] Wang S, Liu X, Chen S, Liu Z, Zhang X, Liang XJ, Li L. Regulation of Ca(2+) signaling for drug-resistant breast cancer therapy with mesoporous silica nanocapsule encapsulated doxorubicin/siRNA cocktail. ACS Nano 2019;13(1):274–83.
[23] Chen S, Shi Y, Zhang X, Ma J. Evaluation of BMP-2 and VEGF loaded 3D printed hydroxyapatite composite scaffolds with enhanced osteogenic capacity in vitro and in vivo. Mater Sci Eng C Mater Biol Appl 2020;112, 110893.
[24] Wu Z, Liang S, Kuai W, Hu L, Qian A. MicroRNAs and long noncoding RNAs: new regulators in cell fate determination of mesenchymal stem cells. RSC Adv 2019;9 (64):37300–11.
[25] Liu M, Zhu H, Wang Y, Sevencan C, Li BL. Functionalized MoS2-based nanomaterials for cancer phototherapy and other biomedical applications. ACS Mater Lett 2021;3(5):462–96.
[26] Kerativitayanan P, Carrow JK, Gaharwar AK. Nanomaterials for engineering stem cell responses. Adv Healthc Mater 2015;4(11):1600–27.
[27] Ko WK, Heo DN, Moon HJ, Lee SJ, Bae MS, Lee JB, Sun IC, Jeon HB, Park HK, Kwon IK. The effect of gold nanoparticle size on osteogenic differentiation of adipose-derived stem cells. J Colloid Interface Sci 2015;438:68–76.
[28] Li J, Li JJ, Zhang J, Wang X, Kawazoe N, Chen G. Gold nanoparticle size and shape influence on osteogenesis of mesenchymal stem cells. Nanoscale 2016;8(15): 7992–8007.
[29] de Barros HR, Piovan L, Sassaki GL, de Araujo Sabry D, Mattoso N, Nunes AM, Meneghetti MR, Riegel-Vidotti IC. Surface interactions of gold nanorods and polysaccharides: from clusters to individual nanoparticles. Carbohydr Polym 2016;152:479–86.
[30] Wang K, Tong X, Lin J, Wei A, Li Y, Dargusch M, Wen C. Binary Zn–Ti alloys for orthopedic applications: corrosion and degradation behaviors, friction and wear performance, and cytotoxicity. J Mater Sci Technol 2021;74:216–29.
[31] Foroutan T. The effects of zinc oxide nanoparticles on differentiation of human mesenchymal stem cells to osteoblast. Nanomed J 2014;1(5):308–14.

[32] Zhou X, Feng W, Qiu K, Chen L, Wang W, Nie W, Mo X, He C. BMP-2 derived peptide and dexamethasone incorporated mesoporous silica nanoparticles for enhanced osteogenic differentiation of bone mesenchymal stem cells. ACS Appl Mater Interfaces 2015;7(29):15777–89.
[33] Mahla RS. Stem cells applications in regenerative medicine and disease therapeutics. Int J Cell Biol 2016;2016:6940283.
[34] Gao C, Lin Z, Wu Z, Lin X, He Q. Stem-cell-membrane camouflaging on near-infrared photoactivated upconversion nanoarchitectures for in vivo remote-controlled photodynamic therapy. ACS Appl Mater Interfaces 2016;8(50):34252–60.
[35] Leader B, Beca QJ, Golan DE. Protein therapeutics: a summary and pharmacological classification. Nat Rev Drug Discov 2008;7(1):21–39.
[36] Martino MM, Briquez PS, Maruyama K, Hubbell JA. Extracellular matrix-inspired growth factor delivery systems for bone regeneration. Adv Drug Deliv Rev 2015;94:41–52.
[37] Bishop GB, Einhorn TA. Current and future clinical applications of bone morphogenetic proteins in orthopaedic trauma surgery. Int Orthop 2007;31(6):721–7.
[38] Davoodi P, Lee LY, Xu Q, Sunil V, Sun Y, Soh S, Wang CH. Drug delivery systems for programmed and on-demand release. Adv Drug Deliv Rev 2018;132:104–38.
[39] Vo TN, Kasper FK, Mikos AG. Strategies for controlled delivery of growth factors and cells for bone regeneration. Adv Drug Deliv Rev 2012;64(12):1292–309.
[40] Yanez-Mo M, Siljander PR, Andreu Z, Zavec AB, Borras FE, Buzas EI, Buzas K, Casal E, Cappello F, Carvalho J, Colas E, Cordeiro-da Silva A, Fais S, Falcon-Perez JM, Ghobrial IM, Giebel B, Gimona M, Graner M, Gursel I, Gursel M, Heegaard NH, Hendrix A, Kierulf P, Kokubun K, Kosanovic M, Kralj-Iglic V, Kramer-Albers EM, Laitinen S, Lasser C, Lener T, Ligeti E, Line A, Lipps G, Llorente A, Lotvall J, Mancek-Keber M, Marcilla A, Mittelbrunn M, Nazarenko I, Nolte-'t Hoen EN, Nyman TA, O'Driscoll L, Olivan M, Oliveira C, Pallinger E, Del Portillo HA, Reventos J, Rigau M, Rohde E, Sammar M, Sanchez-Madrid F, Santarem N, Schallmoser K, Ostenfeld MS, Stoorvogel W, Stukelj R, Van der Grein SG, Vasconcelos MH, Wauben MH, De Wever O. Biological properties of extracellular vesicles and their physiological functions. J Extracell Vesicles 2015;4, 27066.
[41] Hessvik NP, Llorente A. Current knowledge on exosome biogenesis and release. Cell Mol Life Sci 2018;75(2):193–208.
[42] Ludwig AK, Giebel B. Exosomes: small vesicles participating in intercellular communication. Int J Biochem Cell Biol 2012;44(1):11–5.
[43] Colombo M, Raposo G, Thery C. Biogenesis, secretion, and intercellular interactions of exosomes and other extracellular vesicles. Annu Rev Cell Dev Biol 2014;30:255–89.
[44] van Niel G, D'Angelo G, Raposo G. Shedding light on the cell biology of extracellular vesicles. Nat Rev Mol Cell Biol 2018;19(4):213–28.
[45] Maione F, Cappellano G, Bellan M, Raineri D, Chiocchetti A. Chicken-or-egg question: which came first, extracellular vesicles or autoimmune diseases? J Leukoc Biol 2020;108(2):601–16.
[46] Roefs MT, Sluijter JPG, Vader P. Extracellular vesicle-associated proteins in tissue repair. Trends Cell Biol 2020;30(12):990–1013.
[47] Buzas EI, Toth EA, Sodar BW, Szabo-Taylor KE. Molecular interactions at the surface of extracellular vesicles. Semin Immunopathol 2018;40(5):453–64.
[48] Cloutier N, Tan S, Boudreau LH, Cramb C, Subbaiah R, Lahey L, Albert A, Shnayder R, Gobezie R, Nigrovic PA, Farndale RW, Robinson WH, Brisson A, Lee DM, Boilard E. The exposure of autoantigens by microparticles underlies the formation of potent inflammatory components: the microparticle-associated immune complexes. EMBO Mol Med 2013;5(2):235–49.

[49] Raposo G, Nijman HW, Stoorvogel W. B lymphocytes secrete antigen-presenting vesicles. J Exp Med 1996;183(3):1161–72.
[50] Yan HC, Yu TT, Li J, Qiao YQ, Wang LC, Zhang T, Li Q, Zhou YH, Liu DW. The delivery of extracellular vesicles loaded in biomaterial scaffolds for bone regeneration. Front Bioeng Biotechnol 2020;8:1015.
[51] Zhao AG, Shah K, Cromer B, Sumer H. Mesenchymal stem cell-derived extracellular vesicles and their therapeutic potential. Stem Cells Int 2020;2020:8825771.
[52] Mardpour S, Ghanian MH, Sadeghi-Abandansari H, Mardpour S, Nazari A, Shekari F, Baharvand H. Hydrogel-mediated sustained systemic delivery of mesenchymal stem cell-derived extracellular vesicles improves hepatic regeneration in chronic liver failure. ACS Appl Mater Interfaces 2019;11(41):37421–33.
[53] Qin Y, Wang L, Gao Z, Chen G, Zhang C. Bone marrow stromal/stem cell-derived extracellular vesicles regulate osteoblast activity and differentiation in vitro and promote bone regeneration in vivo. Sci Rep 2016;6:21961.
[54] Chen P, Zheng L, Wang Y, Tao M, Xie Z, Xia C, Gu C, Chen J, Qiu P, Mei S, Ning L, Shi Y, Fang C, Fan S, Lin X. Desktop-stereolithography 3D printing of a radially oriented extracellular matrix/mesenchymal stem cell exosome bioink for osteochondral defect regeneration. Theranostics 2019;9(9):2439–59.
[55] Liu X, Yang Y, Li Y, Niu X, Zhao B, Wang Y, Bao C, Xie Z, Lin Q, Zhu L. Integration of stem cell-derived exosomes with in situ hydrogel glue as a promising tissue patch for articular cartilage regeneration. Nanoscale 2017;9(13):4430–8.
[56] Kokaia Z, Martino G, Schwartz M, Lindvall O. Cross-talk between neural stem cells and immune cells: the key to better brain repair? Nat Neurosci 2012;15(8):1078–87.
[57] Wollert KC, Drexler H. Cell therapy for the treatment of coronary heart disease: a critical appraisal. Nat Rev Cardiol 2010;7(4):204–15.
[58] Amado LC, Saliaris AP, Schuleri KH, et al. Cardiac repair with intramyocardial injection of allogeneic mesenchymal stem cells after myocardial infarction. Proc Natl Acad Sci 2005;102(32):11474–9.
[59] Jokerst JV, Khademi C, Gambhir SS. Intracellular aggregation of multimodal silica nanoparticles for ultrasound-guided stem cell implantation. Sci Transl Med 2013;5 (177) [177ra35].
[60] Blanpain C, Fuchs E. Stem cell plasticity. Plasticity of epithelial stem cells in tissue regeneration. Science 2014;344(6189):1242281.
[61] Lane SW, Williams DA, Watt FM. Modulating the stem cell niche for tissue regeneration. Nat Biotechnol 2014;32(8):795–803.
[62] Li Z, Suzuki Y, Huang M, Cao F, Xie X, Connolly AJ, Yang PC, Wu JC. Comparison of reporter gene and iron particle labeling for tracking fate of human embryonic stem cells and differentiated endothelial cells in living subjects. Stem Cells 2008;26 (4):864–73.
[63] Wang H, Cao F, De A, Cao Y, Contag C, Gambhir SS, Wu JC, Chen X. Trafficking mesenchymal stem cell engraftment and differentiation in tumor-bearing mice by bioluminescence imaging. Stem Cells 2009;27(7):1548–58.
[64] Kidd S, Spaeth E, Dembinski JL, Dietrich M, Watson K, Klopp A, Battula VL, Weil M, Andreeff M, Marini FC. Direct evidence of mesenchymal stem cell tropism for tumor and wounding microenvironments using in vivo bioluminescent imaging. Stem Cells 2009;27(10):2614–23.
[65] Zhang Y, Hong G, Zhang Y. Ag2S quantum dot: a bright and biocompatible fluorescent nanoprobe in the second near-infrared window. ACS Nano 2012;6(5):3695–702.
[66] Zhao P, Xu Q, Tao J, Jin Z, Pan Y, Yu C, Yu Z. Near infrared quantum dots in biomedical applications: current status and future perspective. Wiley Interdiscip Rev Nanomed Nanobiotechnol 2018;10(3), e1483.

[67] Duan X, Wang Y, Zhang F, Lu L, Cao M, Lin B, Zhang X, Mao J, Shuai X, Shen J. Superparamagnetic Iron oxide-loaded cationic polymersomes for cellular MR imaging of therapeutic stem cells in stroke. J Biomed Nanotechnol 2016;12(12):2112–24.
[68] Bulte JWM, Zhang SC, Van Gelderen P. Neurotransplantation of magnetically labeled oligodendrocyte progenitors: magnetic resonance tracking of cell migration and myelination. Proc Natl Acad Sci 1999;96(26):15256–61.
[69] Chhour P, Naha PC, O'Neill SM, Litt HI, Reilly MP, Ferrari VA, Cormode DP. Labeling monocytes with gold nanoparticles to track their recruitment in atherosclerosis with computed tomography. Biomaterials 2016;87:93–103.
[70] Comenge J, Fragueiro O, Sharkey J, Taylor A, Held M, Burton NC, Park BK, Wilm B, Murray P, Brust M, Levy R. Preventing plasmon coupling between gold nanorods improves the sensitivity of photoacoustic detection of labeled stem cells in vivo. ACS Nano 2016;10(7):7106–16.
[71] Chen F, Ma M, Wang J, Wang F, Chern SX, Zhao ER, Jhunjhunwala A, Darmadi S, Chen H, Jokerst JV. Exosome-like silica nanoparticles: a novel ultrasound contrast agent for stem cell imaging. Nanoscale 2017;9(1):402–11.
[72] Notingher I, Bisson I, Polak JM, Hench LL. In situ spectroscopic study of nucleic acids in differentiating embryonic stem cells. Vib Spectrosc 2004;35(1–2):199–203.
[73] Notingher I, Bisson I, Bishop AE, et al. In situ spectral monitoring of mRNA translation in embryonic stem cells during differentiation in vitro. Anal Chem 2004;76 (11):3185–93.
[74] Chan JW, Lieu DK, Huser T. Label-free separation of human embryonic stem cells and their cardiac derivatives using Raman spectroscopy. Anal Chem 2009;81(4):1324–31.
[75] Huefner A, Kuan WL, Barker RA, Mahajan S. Intracellular SERS nanoprobes for distinction of different neuronal cell types. Nano Lett 2013;13(6):2463–70.
[76] Shi C, Cao X, Chen X, Sun Z, Xiang Z, Zhao H, Qian W, Han X. Intracellular surface-enhanced Raman scattering probes based on TAT peptide-conjugated Au nanostars for distinguishing the differentiation of lung resident mesenchymal stem cells. Biomaterials 2015;58:10–25.
[77] Culum NM, Cooper TT, Bell GI, Hess DA, Lagugne-Labarthet F. Characterization of extracellular vesicles derived from mesenchymal stromal cells by surface-enhanced Raman spectroscopy. Anal Bioanal Chem 2021;413(20):5013–24.
[78] Milewska A, Sigurjonsson OE, Leosson K. SERS imaging of mesenchymal stromal cell differentiation. ACS Appl Bio Mater 2021;4(6):4999–5007.
[79] Ko J, Ham J, Lee H, Lee K, Koh WG. Integration of a fiber-based cell culture and biosensing system for monitoring of multiple protein markers secreted from stem cells. Biosens Bioelectron 2021;193, 113531.
[80] Sahu A, Jeon J, Lee MS, Yang HS, Tae G. Nanozyme impregnated mesenchymal stem cells for hepatic ischemia-reperfusion injury alleviation. ACS Appl Mater Interfaces 2021;13(22):25649–62.
[81] Kwon HJ, Kim D, Seo K, Kim YG, Han SI, Kang T, Soh M, Hyeon T. Ceria nanoparticle systems for selective scavenging of mitochondrial, intracellular, and extracellular reactive oxygen species in Parkinson's disease. Angew Chem Int Ed Eng 2018;57 (30):9408–12.
[82] Yu D, Ma M, Liu Z, Pi Z, Du X, Ren J, Qu X. MOF-encapsulated nanozyme enhanced siRNA combo: control neural stem cell differentiation and ameliorate cognitive impairments in Alzheimer's disease model. Biomaterials 2020;255, 120160.

CHAPTER 4

The enhanced therapeutic potential of MSC with biomaterials application

Pingping Jia[a,b], Zhikun Guo[b,c], and Zongjin Li[d,e]
[a]College of Medicine, Xinxiang University, Xinxiang, China
[b]Henan Key Laboratory of Medical Tissue Regeneration, Xinxiang Medical University, Xinxiang, China
[c]Department of Cardiology, Zhengzhou Seventh People's Hospital, Zhengzhou, China
[d]Nankai University School of Medicine, Tianjin, China
[e]Tianjin Key Laboratory of Human Development and Reproductive Regulation, Tianjin Central Hospital of Gynecology Obstetrics, Nankai University Affiliated Hospital of Obstetrics and Gynecology, Tianjin, China

4.1 Introduction

Due to the unique characteristics of mesenchymal stem cells (MSCs), including self-renewal, differentiation, ex vivo expansion, and low immunogenicity, MSCs-based therapy has been widely used in tissue engineering and regenerative medicine [1] and is becoming an increasingly attractive approach for a variety of diseases, such as cardiovascular diseases [2], neurological disorders [3], wound repair [4], and so on.

However, there remain limitations to the efficacy of MSCs-based therapy when transplanted into the body because of the changed microenvironment, which leads to poor cell survival and engraftment, as well as impaired function [5,6]. Therefore, improving the retention and homing of transplanted MSCs is of great significance to its therapeutic use in vivo. Biomaterials, which can improve the survival and differentiation of transplanted cells in vivo by simulating the extracellular matrix [7,8], provide a strategy to overcome the limitations and enhance the therapeutic effect of MSCs [8] (Fig. 1). Furthermore, previous studies have proved the elevated therapeutic efficiency of MSCs with biomaterials as delivery scaffolds, such as the RGD-modified self-assembling D-form peptide hydrogel (Nap-^{D}F^{D}FKGRGD) that serves as an artificial niche for MSCs, leading to superior limb salvage in a mouse ischemic hind limb model, as it enhances the therapeutic effects of MSCs for hind limb ischemia [9] (Fig. 2). In this chapter, we will mainly discuss the enhanced therapeutic potential of MSCs with biomaterials.

https://doi.org/10.1016/B978-0-323-95346-7.00002-6

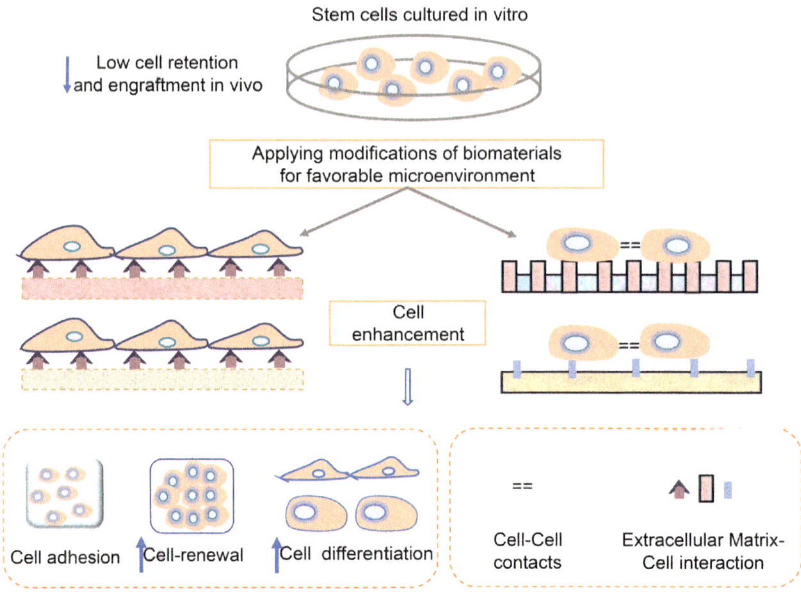

Fig. 1 Biomaterials as the scaffolds to transplant MSCs to improve the survival and differentiation of cells in vivo by providing a favorable microenvironment. *(Reproduced under a Creative Commons Attribution 4.0 International License from Zhao X, Li Q, Guo Z, Li Z, Constructing a cell microenvironment with biomaterial scaffolds for stem cell therapy. Stem Cell Res Ther 2021;12(1):583.)*

4.2 MSCs-based therapy

Nowadays, numerous studies suggest that MSCs-based therapy is a promising approach for many clinical diseases, in which traditional treatment methods can't completely and effectively alleviate the suffering, such as cardiovascular diseases, ischemic diseases, neurological disorders, hepatic disease, etc. In order to promote the repair and regeneration of injured tissue, MSCs and their derivatives [10,11], which possess the ability of self-renewal, multidirectional differentiation, secretion of active factors, and low immunogenicity, are transplanted to damaged areas through blood circulation or direct injection. Comparing with other stem cells, MSCs have tremendous advantages in obtaining, cultivating, and expanding [12], including bone marrow mesenchymal stem cells [13,14], umbilical cord-derived mesenchymal stem cells [15], and placental derived mesenchymal stem cells [16].

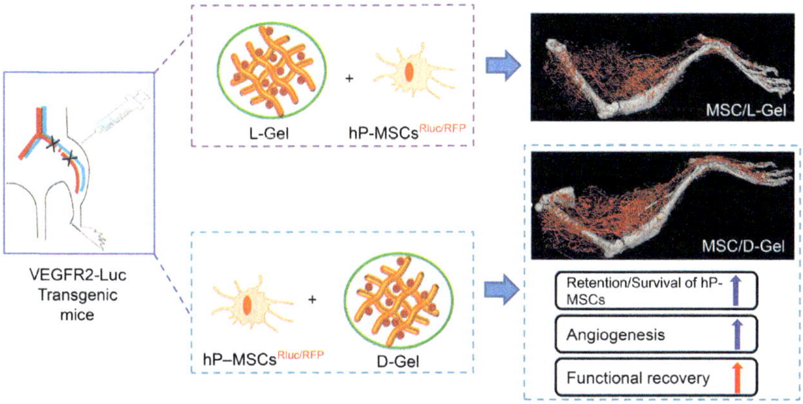

Fig. 2 The RGD-modified self-assembling D-form peptide hydrogel (Nap-DFDFKGRGD) enhances the therapeutic effect of MSCs via simulating the extracellular matrix to serve as an artificial niche. *(Reprinted with copyright permission from Jia P, Zhao X, Liu Y, Liu M, Zhang Q, Chen S, Huang H, Jia Y, Chang Y, Han Z, Han ZC, Li Q, Guo Z, Li L, The RGD-modified self-assembling D-form peptide hydrogel enhances the therapeutic effects of mesenchymal stem cells (MSC) for hindlimb ischemia by promoting angiogenesis, Chem Eng J 2022;450:138004.)*

4.3 Challenges for MSCs-based therapy

While a significant body of research has shown that MSCs-based therapy is a promising strategy, which has been widely used in tissue engineering and regenerative medicine, there still remain challenges. On the one hand, transplanted cells are easily washed away by the blood flow or flow out from the tissue [17,18]; for example, when the MSCs are transplanted into ischemic region of a myocardial infarction animal model [19], only a small number of cells are retained, which greatly reduces the therapeutic effects. On the other hand, due to the change of microenvironment, for example, to ischemic or hypoxic, MSCs have performed poorly in survival, proliferation, and differentiation, and thus shown limited therapeutic efficiency [6].

Biomaterials play a key role in tissue engineering and regenerative medicine. A large number of studies have shown that biomaterials can simulate the extracellular matrix, provide a suitable microenvironment, and protect cells from ischemia-hypoxia as well as other damage [8,20]. Based on these findings, biomaterials can be used as scaffolds to deliver MSCs, enhancing transplanted cellular engraftment, promoting cell differentiation and paracrine signaling, etc. Hence, the combination of biomaterials and MSCs has tremendous potential to improve the therapeutic effects of MSCs.

4.4 Characteristics of biomaterials

Biomaterials, as the scaffold for MSCs, must have some essential characteristics that mostly involve biocompatibility and biodegradability [21]. Biocompatibility means that biomaterials have no toxic or injurious influence on biological systems [22,23]. Under this premise, biomaterials can support cellular activity, provide the environment favor for cell growth and proliferation, as well as being conducive to tissue repair. Furthermore, biomaterial should be biodegradable; this event occurs over a period of time after biomaterials co-transplantation with MSCs in vivo, and the degradation products should have no toxic effect and produce no inflammatory responses in the organism [24]. In addition, a crucial aspect of biomaterials is that they can also be modified with additional functional molecules to enhance the therapeutic effects of cells, such as transforming growth factor β(TGF-β), which assist in reducing immune response [25], and insulin-like growth factor (IGF-1), which promotes cell proliferation and tissue repair [6] (Fig. 3).

4.5 Classification of biomaterials

According to their origin, biomaterials are generally classified as natural or synthetic [26].

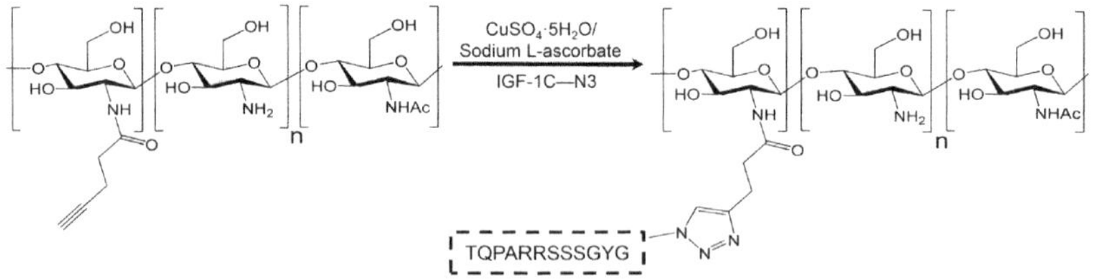

Fig. 3 IGF-1C is grafted onto chitosan by a click reaction, aiming to develop a chitosan (CS)-based injectable hydrogel with immobilized IGF-1C (CS–IGF-1C) to strengthen the supportive niche function, favoring survival and therapeutic benefits of mesenchymal stem cells. *(Reprinted with copyright permission from Zhao N, Yue Z, Cui J, Yao Y, Song X, Cui B, Qi X, Han Z, Han ZC, Guo Z, He ZX, Li Z, IGF-1C domain-modified hydrogel enhances therapeutic potential of mesenchymal stem cells for hindlimb ischemia, Stem Cell Res Ther 2019;10(1):129.)*

4.5.1 Natural biomaterials

Natural biomaterials have similar biological characteristics to the extracellular matrix (ECM) and are highly biocompatibility, making them better for cell proliferation and differentiation. But at the same time, they also have some disadvantages because of these properties. For example, natural biomaterials are more likely to be degraded by protease, compared to their synthetic counterparts, which means a shorter retention time in vivo. Furthermore, natural biomaterials are difficult to purify, as well as the quality not being easy to control. The ECM, which is mainly composed of collagen, proteoglycans, glycoproteins, and glycosaminoglycans, is a highly complex three-dimensional macromolecular network structure that not only provides support to cells but also plays key regulatory roles in many cellular process involved in growth, migration, and differentiation [27–29]. Natural biomaterials, including hyaluronic acid, fibrin, artificial matrix glue, and chitosan, have been widely used in many diseases, such as chitosan application in dentistry [30], and fibrin-based biomaterials used in cardiac patches [31].

Hyaluronic acid (HA), a nonsulfated glycosaminoglycan (GAG), is a major component of the extracellular matrix (ECM) and plays a key role in wound healing and tissue repair processes owing to its capacity for cell signaling, cell proliferation, and anti-inflammatory response [32]. Due to the characteristics of photo-crosslinking, HA can be used to deliver stem cells and improve cell viability. As a highly hydrated polyanionic macromolecule and its glycosaminoglycan properties, HA also has an angiogenic effect related to its molecular weight. HA, which possesses low molecular weight (100,000 Da in serum), accelerates the proliferation and migration of endothelial cells and is conducive to angiogenesis, whereas high molecular weight (8 million Da in vitreous) inhibits angiogenesis [33]. Nowadays, HA is widely used to produce wound dressings, such as hyaluronic acid sponges containing arginine and epidermal growth factor [34], and a hyaluronic acid-based hydrogel with miRNA-laden nanoparticles for chronic diabetic wound treatment via antiinflammatory and angiogenesis mechanisms [35].

Fibrin, a protein polymer, is essential to the physiological blood coagulation cascade [36,37]. In addition to its important role in hemostasis, fibrin also has the unique physiological and structural advantage of a porous nature, therefore, this natural hydrogel possesses excellent viscoelasticity and deformability despite its low protein content [38]. At present, fibrin is being used as an excellent cell delivery scaffold by increasing proliferation and differentiation; for example, the system of keratinocytes and fibrin is beneficial for the healing of chronic skin wounds. Simultaneously, it can also act as a release system for vascular endothelial growth factor (VEGF) and basic fibroblast growth factor (bFGF) [39]. Moreover, fibrin can combine with other natural or synthetic biomaterials to improve its degradation rate and mechanical strength, as on its own it has easily degraded properties [38].

Matrigel, a soluble protein, was developed from the Engelbreth-Holm-Swarm (EHS) murine tumor, which produces a basement-membrane matrix [40]. The main components of Matrigel include laminin, collagen, entactin, and heparin sulfate. Laminins are indispensable building blocks for cellular networks and transmit signals controlling cellular behavior [41]. Collagens not only regulate the proliferation, migration, and differentiation of cells by the receptor families, but can also contribute to mechanical properties, organization, and shape of tissues [42]. Entactin has calcium-binding characteristics that assist in the matrix assembly process [43]. Furthermore, Matrigel contains various growth factors, including transforming growth factor beta (TGF-beta), epidermal growth factor (EGF), insulin-like growth factor 1, bovine fibroblast growth factor (bFGF), and platelet-derived growth factor (PDGF) [44]. Matrigel has been shown to support two- or three-dimensional cell cultivation, such as the Matrigel and endothelial cells co-transplanted platform, in which the retention rate of endothelial cells was significantly increased after transplanting into a mouse myocardial ischemia model [45].

Chitosan, a natural polycationic linear polysaccharide composed of β-(1–4)-linked D-glucosamine and *N*-acetyl-D-glucosamine, is a derivative of chitin [46]. Chitosan is insoluble in aqueous and alkaline solutions; therefore, the modification of its structure is prerequisite to improve its physicochemical properties [47]. Recently, chitosan has been widely used in wound healing, drug transport, and tissue engineering due to its biocompatible, biodegradable, antitumor, and antimicrobial characteristics [48–50].

4.5.2 Synthetic biomaterials

Compared to natural biomaterials, synthetic biomaterials have several disadvantages, including insufficient ECM-mimicry, immunoreaction due to possible toxic residual solvents after degradation, and the uncontrollable degradation rate [51]. Hence, in order to manage the degradation rate, mechanical properties, and low immunogenicity, synthetic biomaterials need to be artificially modified [52]. Among these biomaterials, biometal materials, bioceramic materials, polymer materials, etc., are widely used.

Biometal materials are frequently bioinert materials, which exhibit excellent heat, corrosion, wear, and chemical resistance, and high mechanical strength [53]. The main biomedical materials that have been used are titanium and titanium-based alloys, stainless steel, cobalt-based alloys, etc. [54–56]. Titanium alloy materials are commonly used in surgical implants, and production and manufacture of orthopedic devices, such as artificial joints, dental implants, and vascular stents [57–59]. Stainless steel is used extensively in stomatology and in internal fracture fixation devices and artificial joints [55].

Bioceramic materials, which include bioabsorbent ceramics, bioactive ceramics, and bioinert ceramics, have similar performance to bone materials, such as surface structure, friction coefficient, density, and thermal conductivity [60]. Absorbable bioceramic material refers to materials that can be removed within the organism through asymptotic degradation, participating in metabolism, or being absorbed by tissues, such as calcium polyphosphate, tricalcium α-phosphate, tricalcium β-phosphate, and porous hydroxyapatite [61]. Bioactive ceramics are a kind of material that benefits tissue repair, replacement, and regeneration due to superior biocompatibility as well as being nondegradable in vivo [62]. Bioinert ceramics, including alumina ceramics and oxidation zirconium ceramics, are mainly used for bone repair, the repair and replacement of arthrosis, artificial tooth roots, middle ear ossicles, and heart valves [63].

Saturated aliphatic polyesters such as polyglycolic acid (PGA), polylactic acid (PLA), and polylactic-polyglycolic acid copolymer (PLGA) are the most commonly used biosynthetic materials [52]. They can make use of ester bonds to carry out de-esterification hydrolysis, and their degradation products, glycolipid and lactic acid, can be eliminated in vivo through metabolic pathways, so they have good biocompatibility and biodegradability [64].

PGA is a hydrophilic, highly crystalline polymer that provides flexibility in adjusting the pore size and flexibility of the material, but degrades at a faster rate. PGA tablets combined with natural biomaterial fibrin glue can promote wound healing during oral surgery.

PLA is structurally similar to PGA, but contains additional methyl groups, so PLA biomaterials exhibit excellent mechanical properties as scaffolds, making them suitable for weight-bearing applications such as orthopedic fixation devices.

PLGA, polylactic acid polyglycolic acid copolymer, can achieve a degradation rate between that of PLA and PGA by changing the proportion between the monomers [65]. At present, in order to increase the bionic property of PLGA, it is often combined with other materials and modified.

Polyanhydride can be degraded into nontoxic diacid by-products in vivo and discharged as metabolites, and has good biocompatibility. At the same time, it is also hydrophobic, so that when the surface of the biomaterial is eroded, it will not degrade as a whole but will degrade locally, thus protecting the drug or cell wrapped in it [66].

Polyurethanes, usually synthesized by polycondensation of diisocyanates with alcohols/amines, have shown widespread versatility in terms of mechanical properties, biological properties, physical properties, biodegradability, and blood and tissue compatibility due to their segmented structural properties.

Polyphosphazenes are a relatively new class of inorganic-organic hybrid polymers. The main chain of the polyphosphazenes enables the biomaterials to have good flexibility, and the degradation and hydrophilicity/hydrophobicity of the materials can be controlled through modification of pendant groups [67].

4.6 Three-dimensional culture of MSCs with biomaterials

When MSCs are transplanted in vivo, they often have different characteristics from in vitro cultures; this is mainly because in vitro culturing of MSCs is conducted in a flat 2D cell system that is attached to culture bottles and petri dishes, where the single-layer 2D environment provides cells with nutrients and growth factors. The living environment of MSCs in vivo is a three-dimensional structure formed by extracellular matrix. The traditional 2D cell culture system is very valuable for cell research, but it cannot truly simulate the endogenous microenvironment. The biomaterial scaffold can simulate the extracellular matrix to form a 3D culture system, provide a real three-dimensional environment for MSCs, and overcome the limitation of pure 2D culture [68]. In the 3D culture system, cells are usually arranged and grown in the form of cell-cell or cell-scaffold, which more accurately simulates the microenvironment in which the cells live in the body and is more conducive to cell growth, migration, proliferation, and differentiation. Therefore, 3D cell culture has become a necessary method for the research of MSCs systems [69,70] (Fig. 4). Nowadays, a large number of studies have

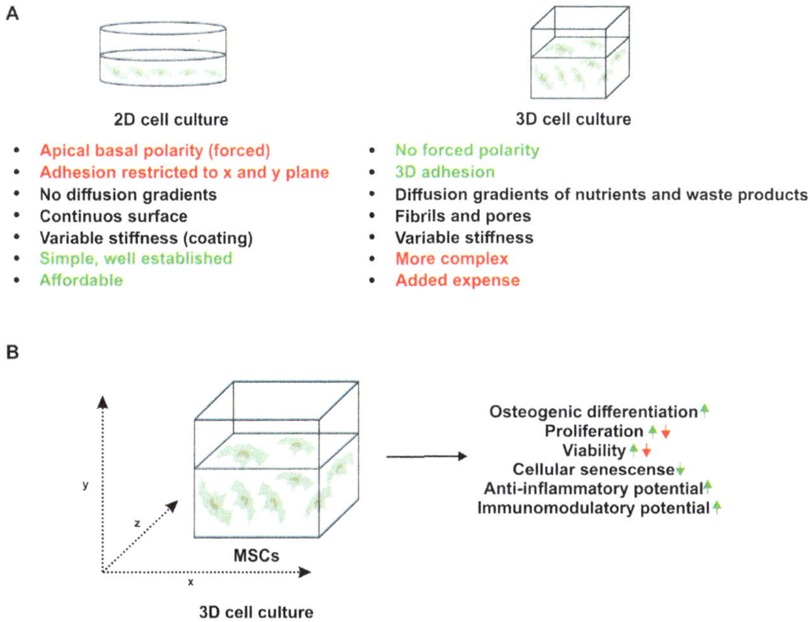

Fig. 4 Effects of 3D cell culture on MSCs. (A) Advantages (marked in *green*) of 3D cell culture compared to traditional 2D cell cultivation. (B) 3D cell culture enhances the osteogenic differentiation potential of MSCs and reduces the cell aging. *(Reprinted with copyright permission from Bicer M, Cottrell GS, Widera D, Impact of 3D cell culture on bone regeneration potential of mesenchymal stromal cells, Stem Cell Res Ther 2021;12(1):31.)*

shown that the proliferation and differentiation of cells in the 3D culture system are superior to that of the 2D culture system [71,72].

4.7 Therapeutic potential of MSC with biomaterials

4.7.1 Spinal cord injury

The loss of neurons and the degeneration of axons after spinal cord injury lead to impaired sensory and motor function, so the theoretical method for treating this disease is to regenerate the damaged axons through the injury site [73,74]. MSCs can improve damaged spinal cord by providing nutritional support, promoting autologous regeneration, or directly replacing neurons and their supporting cells, thereby promoting functional recovery. However, a large number of glial scars and inflammatory cytokines are generated at the site of spinal cord injury, which hinder the regeneration of neuronal axons to synaptic targets [75,76]. A biomaterial scaffold can serve as a bridge to guide the regenerated axons of stem cells to enter the target tissue through the injury site, thereby promoting tissue repair and functional recovery. The scaffold should be a continuous channel structure similar to the ECM, which can help axonal regeneration and guide the growth of nerve structures [77]. MSCs can differentiate into osteoblasts, chondrocytes, and adipocytes [78]. More and more studies have shown that the combination of biological material scaffold and MSCs can provide a potential treatment for the repair of spinal cord injury [79], such as the NeuroRegen scaffolds, a collagen scaffold; the recovery of sensory and motor functions was observed in the treatment of two patients when this biomaterial was co-transplanted with MSCs [80].

4.7.2 Diabetic foot

There are two types of diabetes, Type 1 (T1D) and Type 2 (T2D). Type 1 diabetes mellitus (T1D) refers to the fact that autoimmune deficiency leads to the destruction and function impairment of islet β cells, further resulting in a large decrease in insulin secretion, and absolute lack of insulin [81]. Type 2 diabetes mellitus (T2D) is mainly because of islet β cell dysfunction or the target organs, such liver and muscle, are resistant to the effects of insulin [82]. The treatment of T1D mainly depends on exogenous insulin injection, while that of T2D is mainly oral hypoglycemic drugs.

Diabetes eventually causes a variety of complications, such as diabetic foot. Mesenchymal stem cells, such as bone marrow mesenchymal stem cells and adipose-derived mesenchymal stem cells, can reduce inflammatory response and secrete a large number of antiinflammatory factors, including IL-6, TGF-β, and E2, and also secrete angiogenic factor VEGF, etc., to promote the regeneration and repair of damaged tissues [83]. This can also be used for the treatment of diabetic foot. However, MSCs transplantation alone cannot achieve the ideal therapeutic effect. In order to improve the curative effect, biological materials can be used as scaffolds to transfer stem cells. Hyaluronic acid can promote angiogenesis, so the co-transplantation of MSCs and hyaluronic acid-based spongy hydrogel can rapidly promote the healing of diabetic foot ulcers [84].

4.7.3 Myocardial infarction

Partial or complete occlusion of the coronary artery causes severe ischemia and hypoxia of the myocardial tissue, leading to myocardial infarction.

Ventricular remodeling, inflammatory cell infiltration and fibroblast proliferation, fibrous scar tissue formation, ventricular remodeling, and eventually, ventricular failure occur after myocardial infarction [85,86]. In recent years, some results have been achieved in using MSCs to improve cardiac function and reduce ventricular remodeling in cardiovascular diseases [87].

To date, methods of delivering stem cells to the heart have included intravenous (IV), intracoronary (IC), or direct epicardial injection. The method of intravenously transferring cells is the least invasive, but most cells are trapped in the lung and less than 1% return to the heart. Cells can be injected directly into the infarct site via the coronary artery, but there are still large numbers of cells that are lost, either because of compression or because they are washed away by blood, and finally only a small number of stem cells are left in the heart. Compared with these two methods, more cells directly injected by intramuscular injection are transplanted to the site of heart injury. Therefore, the low survival rate and the retention rate of the cell transplanted into the infarct myocardium reduce the therapeutic effect of the MSCs, although there are multiple pathways through which the cells can be delivered.

A biomaterial scaffold can form a three-dimensional environment similar to that in vivo, which is the key for improving the treatment effect of cardiovascular system stem cells. For example, the co-transplantation of MSCs and natural biomaterial fibrin hydrogel into the left ventricular wall of infarcted myocardium in rats significantly increased the survival time and number of MSCs in vivo. The biomaterial scaffold for the cardiovascular system has a porous 3D structure that can provide a basis for vascularization and myolization. In addition, a biomaterial scaffold with the ability to release NO can increase the oxygen level in the infarct area, thereby increasing the proliferation and survival of MSCs [11].

4.7.4 Wound repair

A wound is formed due to damage to the normal structure of skin tissue. The traditional treatment methods include debridement and suture, infection control, etc., but their efficacy for healing of complex wounds is limited. MSCs have the potential of self-renewal and multidirectional differentiation, and can also secrete bioactive factors and immunoregulatory factors, including vascular endothelial growth factor (VEGF), epidermal growth factor (EGF), keratinocyte growth factor (KGF), insulin-like growth factor (IGF), tumor necrosis factor (TNF-α), and interleukin-1(IL-1α and

IL-1β), which are conducive to tissue repair and thus promote wound healing [88]. However, after stem cell transplantation, its survival rate and paracrine activity in vivo are decreased, affecting its therapeutic effect.

Biomaterials can be used as scaffolds for the delivery of stem cells to enhance their efficacy. Common natural biological materials include collagen and fibrin. Collagen scaffold and MSCs were transplanted in a diabetic rabbit ear ulcer wound model, and the results showed that the wound healing rate and closure rate were significantly improved. Keratinocytes and MSCs with fibrin gel not only promote wound healing but also increase vascular density and collagen deposition. With regard to synthetic biomaterials, a study showed that the transplantation of poly (N-isopropylacrylamide) (PNIPAM) and polyamidoamine (PAA), namely PNIPAM-PAA hydrogel, with MSCs to ulcers in diabetic mice promoted angiogenesis and wound contraction, and the ulcers healed in a short time [89].

4.7.5 Liver failure

Although the liver has a strong capacity for regeneration, severe liver cell injury or dysfunction can still lead to liver failure, with serious consequences. The most effective treatment for this disease is liver transplantation. However, transplant patients also face complications such as limited clinical donors and immunologic rejection after transplantation. In addition to activating and accelerating the self-healing process of the host liver through paracrine factors, MSCs have the capability of multidirectional differentiation [90]. Therefore, MSCs therapy has emerged as a potential strategy for the treatment of liver failure.

Although traditional cell therapies have shown promising results, problems of poor cell viability and low viability have arisen after direct MSCs transplantation. Biomaterial scaffolds can provide an appropriate microenvironment for transplanted cells to promote cell migration, proliferation, and differentiation. These biomaterials may be natural or synthetic or may take different forms such as nanofiber and porous scaffolds [91,92]. A recent study showed that a kind of 3D PLGA scaffold improves differentiation and function of MSCs into hepatocytes. The results of PCR, WB, and CLSM indicate the hepatogenesis in this collagen-coated poly(lactic-*co*-glycolic acid) (C-PLGA) 3D scaffold [93]. Compared with other biomaterial scaffolds, the acellular liver tissue scaffold retains the natural liver structure and components, and has incomparable advantages in natural components and ultrastructure, thus being a relatively ideal scaffold for transferring cells [94].

4.8 Conclusion

At present, MSCs-based therapy has made some progress and achieved some success in tissue engineering and regenerative medicine. However, due to the changes of microenvironment after cell transplantation, the ideal efficacy cannot be achieved. Biomaterials can be used as cell delivery scaffolds to promote the survival, proliferation, and differentiation of stem cells after transplantation by simulating the extracellular matrix (ECM). Therefore, the combination of biomaterials and MSCs has promoted the wide application of stem cell therapy. There are two types of biological materials, natural and synthetic. Each biological material has its own characteristics, but their common features are biocompatibility and biodegradability. Although this multidisciplinary approach of biomaterials and MSCs offers a promising and potential therapeutic strategy, its clinical transformation still requires more research.

Acknowledgements

This study was financially supported by the National Natural Science Foundation of China (U2004126), the Tianjin Natural Science Foundation (22JCZXJC00170, 21JCZDJC00070), and Tianjin Key Medical Discipline (Specialty) Construction Project (No. TJYXZDXK-043A).

References

[1] Giordano A, Galderisi U, Marino IR. From the laboratory bench to the patient's bedside: an update on clinical trials with mesenchymal stem cells. J Cell Physiol 2007;211(1):27–35.
[2] Bunnell BA, Deng W, Robinson CM, Waldron PR, Bivalacqua TJ, Baber SR, Hyman AL, Kadowitz PJ. Potential application for mesenchymal stem cells in the treatment of cardiovascular diseases. Can J Physiol Pharmacol 2005;83(7):529–39.
[3] Andrzejewska A, Dabrowska S, Lukomska B, Janowski M. Mesenchymal stem cells for neurological disorders. Adv Sci 2021;8(7):2002944.
[4] Guillamat-Prats R. The role of MSC in wound healing, scarring and regeneration. Cell 2021;10(7):1729.
[5] Yin JQ, Zhu J, Ankrum JA. Manufacturing of primed mesenchymal stromal cells for therapy. Nat Biomed Eng 2019;3(2):90–104.
[6] Zhao N, Yue Z, Cui J, Yao Y, Song X, Cui B, Qi X, Han Z, Han ZC, Guo Z, He ZX, Li Z. IGF-1C domain-modified hydrogel enhances therapeutic potential of mesenchymal stem cells for hindlimb ischemia. Stem Cell Res Ther 2019;10(1):129.
[7] Matson JB, Stupp SI. Self-assembling peptide scaffolds for regenerative medicine. Chem Commun (Camb) 2012;48(1):26–33.
[8] Zhao X, Li Q, Guo Z, Li Z. Constructing a cell microenvironment with biomaterial scaffolds for stem cell therapy. Stem Cell Res Ther 2021;12(1):583.
[9] Jia P, Zhao X, Liu Y, Liu M, Zhang Q, Chen S, Huang H, Jia Y, Chang Y, Han Z, Han Z-C, Li Q, Guo Z, Li Z. The RGD-modified self-assembling D-form peptide hydrogel

enhances the therapeutic effects of mesenchymal stem cells (MSC) for hindlimb ischemia by promoting angiogenesis. Chem Eng J 2022;450, 138004.
[10] Zhang K, Zhao X, Chen X, Wei Y, Du W, Wang Y, Liu L, Zhao W, Han Z, Kong D, Zhao Q, Guo Z, Han Z, Liu N, Ma F, Li Z. Enhanced therapeutic effects of mesenchymal stem cell-derived exosomes with an injectable hydrogel for hindlimb ischemia treatment. ACS Appl Mater Interfaces 2018;10(36):30081–91.
[11] Zhang K, Chen X, Li H, Feng G, Nie Y, Wei Y, Li N, Han Z, Han ZC, Kong D, Guo Z, Zhao Q, Li Z. A nitric oxide-releasing hydrogel for enhancing the therapeutic effects of mesenchymal stem cell therapy for hindlimb ischemia. Acta Biomater 2020;113:289–304.
[12] Watanabe Y, Tsuchiya A, Terai S. The development of mesenchymal stem cell therapy in the present, and the perspective of cell-free therapy in the future. Clin Mol Hepatol 2021;27(1):70–80.
[13] Augello A, De Bari C. The regulation of differentiation in mesenchymal stem cells. Hum Gene Ther 2010;21(10):1226–38.
[14] Minteer D, Marra KG, Rubin JP. Adipose-derived mesenchymal stem cells: biology and potential applications. Adv Biochem Eng Biotechnol 2013;129:59–71.
[15] Fan CG, Zhang QJ, Zhou JR. Therapeutic potentials of mesenchymal stem cells derived from human umbilical cord. Stem Cell Rev Rep 2011;7(1):195–207.
[16] Mathew SA, Naik C, Cahill PA, Bhonde RR. Placental mesenchymal stromal cells as an alternative tool for therapeutic angiogenesis. Cell Mol Life Sci 2020;77(2):253–65.
[17] Hashemzadeh MR, Taghavizadeh Yazdi ME, Amiri MS, Mousavi SH. Stem cell therapy in the heart: biomaterials as a key route. Tissue Cell 2021;71, 101504.
[18] Kim H, Kim SL, Choi YH, Ahn YH, Hwang NS. Biomaterials for stem cell therapy for cardiac disease. Adv Exp Med Biol 2018;1064:181–93.
[19] Miao C, Lei M, Hu W, Han S, Wang Q. A brief review: the therapeutic potential of bone marrow mesenchymal stem cells in myocardial infarction. Stem Cell Res Ther 2017;8(1):242.
[20] Velmurugan BK, Bharathi Priya L, Poornima P, Lee LJ, Baskaran R. Biomaterial aided differentiation and maturation of induced pluripotent stem cells. J Cell Physiol 2019;234 (6):8443–54.
[21] Ribeiro JCV, Vieira RS, Melo IM, Araujo VMA, Lima V. Versatility of chitosan-based biomaterials and their use as scaffolds for tissue regeneration. Sci World J 2017;2017: 8639898.
[22] Miranda SC, Silva GA, Hell RC, Martins MD, Alves JB, Goes AM. Three-dimensional culture of rat BMMSCs in a porous chitosan-gelatin scaffold: a promising association for bone tissue engineering in oral reconstruction. Arch Oral Biol 2011;56(1):1–15.
[23] Keong LC, Halim AS. In vitro models in biocompatibility assessment for biomedical-grade chitosan derivatives in wound management. Int J Mol Sci 2009;10(3):1300–13.
[24] Shafiq M, Ali O, Han SB, Kim DH. Mechanobiological strategies to enhance stem cell functionality for regenerative medicine and tissue engineering. Front Cell Dev Biol 2021;9, 747398.
[25] Re'em T, Kaminer-Israeli Y, Ruvinov E, Cohen S. Chondrogenesis of hMSC in affinity-bound TGF-beta scaffolds. Biomaterials 2012;33(3):751–61.
[26] Marin E, Boschetto F, Pezzotti G. Biomaterials and biocompatibility: an historical overview. J Biomed Mater Res A 2020;108(8):1617–33.
[27] Theocharis AD, Skandalis SS, Gialeli C, Karamanos NK. Extracellular matrix structure. Adv Drug Deliv Rev 2016;97:4–27.
[28] Karamanos NK, Theocharis AD, Piperigkou Z, Manou D, Passi A, Skandalis SS, Vynios DH, Orian-Rousseau V, Ricard-Blum S, Schmelzer CEH, Duca L, Durbeej M, Afratis NA, Troeberg L, Franchi M, Masola V, Onisto M. A guide to the composition and functions of the extracellular matrix. FEBS J 2021;288(24):6850–912.

[29] Frantz C, Stewart KM, Weaver VM. The extracellular matrix at a glance. J Cell Sci 2010;123(Pt 24):4195–200.
[30] Fakhri E, Eslami H, Maroufi P, Pakdel F, Taghizadeh S, Ganbarov K, Yousefi M, Tanomand A, Yousefi B, Mahmoudi S, Kafil HS. Chitosan biomaterials application in dentistry. Int J Biol Macromol 2020;162:956–74.
[31] Barsotti MC, Felice F, Balbarini A, Di Stefano R. Fibrin as a scaffold for cardiac tissue engineering. Biotechnol Appl Biochem 2011;58(5):301–10.
[32] Toole BP. Hyaluronan in morphogenesis. Semin Cell Dev Biol 2001;12(2):79–87.
[33] Burdick JA, Prestwich GD. Hyaluronic acid hydrogels for biomedical applications. Adv Mater 2011;23(12):H41–56.
[34] Matsumoto Y, Kuroyanagi Y. Development of a wound dressing composed of hyaluronic acid sponge containing arginine and epidermal growth factor. J Biomater Sci Polym Ed 2010;21(6–7):715–26.
[35] Yang L, Zhang L, Hu J, Wang W, Liu X. Promote anti-inflammatory and angiogenesis using a hyaluronic acid-based hydrogel with miRNA-laden nanoparticles for chronic diabetic wound treatment. Int J Biol Macromol 2021;166:166–78.
[36] Heher P, Mühleder S, Mittermayr R, Redl H, Slezak P. Fibrin-based delivery strategies for acute and chronic wound healing. Adv Drug Deliv Rev 2018;129:134–47.
[37] Brown AC, Barker TH. Fibrin-based biomaterials: modulation of macroscopic properties through rational design at the molecular level. Acta Biomater 2014;10 (4):1502–14.
[38] Rajangam T, An SS. Fibrinogen and fibrin based micro and nano scaffolds incorporated with drugs, proteins, cells and genes for therapeutic biomedical applications. Int J Nanomedicine 2013;8:3641–62.
[39] de la Puente P, Ludena D, Lopez M, Ramos J, Iglesias J. Differentiation within autologous fibrin scaffolds of porcine dermal cells with the mesenchymal stem cell phenotype. Exp Cell Res 2013;319(3):144–52.
[40] Aisenbrey EA, Murphy WL. Synthetic alternatives to Matrigel. Nat Rev Mater 2020;5 (7):539–51.
[41] Aumailley M. The laminin family. Cell Adhes Migr 2013;7(1):48–55.
[42] Ricard-Blum S. The collagen family. Cold Spring Harb Perspect Biol 2011;3(1), a004978.
[43] Chung AE, Durkin ME. Entactin: structure and function. Am J Respir Cell Mol Biol 1990;3(4):275–82.
[44] Vukicevic S, Kleinman HK, Luyten FP, Roberts AB, Roche NS, Reddi AH. Identification of multiple active growth factors in basement membrane Matrigel suggests caution in interpretation of cellular activity related to extracellular matrix components. Exp Cell Res 1992;202(1):1–8.
[45] He N, Xu Y, Du W, Qi X, Liang L, Wang Y, Feng G, Fan Y, Han Z, Kong D, Cheng Z, Wu JC, He Z, Li Z. Extracellular matrix can recover the downregulation of adhesion molecules after cell detachment and enhance endothelial cell engraftment. Sci Rep 2015;5:10902.
[46] Cheung RC, Ng TB, Wong JH, Chan WY. Chitosan: an update on potential biomedical and pharmaceutical applications. Mar Drugs 2015;13(8):5156–86.
[47] Paul P, Kolesinska B, Sujka W. Chitosan and its derivatives - biomaterials with diverse biological activity for manifold applications. Mini-Rev Med Chem 2019;19(9):737–50.
[48] Kumar MN, Muzzarelli RA, Muzzarelli C, Sashiwa H, Domb AJ. Chitosan chemistry and pharmaceutical perspectives. Chem Rev 2004;104(12):6017–84.
[49] Karagozlu MZ, Kim SK. Anticancer effects of chitin and chitosan derivatives. Adv Food Nutr Res 2014;72:215–25.

[50] Martins AF, Facchi SP, Follmann HD, Pereira AG, Rubira AF, Muniz EC. Antimicrobial activity of chitosan derivatives containing N-quaternized moieties in its backbone: a review. Int J Mol Sci 2014;15(11):20800–32.
[51] Yazdanian M, Alam M, Abbasi K, Rahbar M, Farjood A, Tahmasebi E, Tebyaniyan H, Ranjbar R, Hesam Arefi A. Synthetic materials in craniofacial regenerative medicine: a comprehensive overview. Front Bioeng Biotechnol 2022;10, 987195.
[52] Fournier E, Passirani C, Montero-Menei CN, Benoit JP. Biocompatibility of implantable synthetic polymeric drug carriers: focus on brain biocompatibility. Biomaterials 2003;24(19):3311–31.
[53] Koguchi R, Jankova K, Tanaka M. Fluorine-containing bio-inert polymers: roles of intermediate water. Acta Biomater 2022;138:34–56.
[54] Kaur M, Singh K. Review on titanium and titanium based alloys as biomaterials for orthopaedic applications. Mater Sci Eng C Mater Biol Appl 2019;102:844–62.
[55] Disegi JA, Eschbach L. Stainless steel in bone surgery. Injury 2000;31(Suppl 4):2–6.
[56] Marti A. Cobalt-base alloys used in bone surgery. Injury 2000;31(Suppl 4):18–21.
[57] Xu J, Weng XJ, Wang X, Huang JZ, Zhang C, Muhammad H, Ma X, Liao QD. Potential use of porous titanium-niobium alloy in orthopedic implants: preparation and experimental study of its biocompatibility in vitro. PLoS One 2013;8(11), e79289.
[58] Zhang Y, Gulati K, Li Z, Di P, Liu Y. Dental implant nano-engineering: advances, limitations and future directions. Nanomaterials 2021;11(10):2489.
[59] Gabriel M, Chopra A, Schmidt F. Combined endothelialization promoting and surface binding chimeric conjugate with low thrombogenicity. Bioconjug Chem 2021;32 (8):1602–5.
[60] Sanz JL, Rodríguez-Lozano FJ, Llena C, Sauro S, Forner L. Bioactivity of bioceramic materials used in the dentin-pulp complex therapy: a systematic review. Materials 2019;12(7):1015.
[61] Langstaff S, Sayer M, Smith TJ, Pugh SM, Hesp SA, Thompson WT. Resorbable bioceramics based on stabilized calcium phosphates. Part I: rational design, sample preparation and material characterization. Biomaterials 1999;20(18):1727–41.
[62] Liu X, Morra M, Carpi A, Li B. Bioactive calcium silicate ceramics and coatings. Biomed Pharmacother 2008;62(8):526–9.
[63] Buj-Corral I, Tejo-Otero A. 3D printing of bioinert oxide ceramics for medical applications. J Funct Biomater 2022;13(3):155.
[64] Cameron DJ, Shaver MP. Aliphatic polyester polymer stars: synthesis, properties and applications in biomedicine and nanotechnology. Chem Soc Rev 2011;40(3):1761–76.
[65] Felix Lanao RP, Jonker AM, Wolke JG, Jansen JA, van Hest JC, Leeuwenburgh SC. Physicochemical properties and applications of poly(lactic-co-glycolic acid) for use in bone regeneration. Tissue Eng B Rev 2013;19(4):380–90.
[66] Pan Z, Ding J. Poly(lactide-co-glycolide) porous scaffolds for tissue engineering and regenerative medicine. Interface Focus 2012;2(3):366–77.
[67] Nair LS, Laurencin CT. Polymers as biomaterials for tissue engineering and controlled drug delivery. Adv Biochem Eng Biotechnol 2006;102:47–90.
[68] Baharvand H, Hashemi SM, Kazemi Ashtiani S, Farrokhi A. Differentiation of human embryonic stem cells into hepatocytes in 2D and 3D culture systems in vitro. Int J Dev Biol 2006;50(7):645–52.
[69] Justice BA, Badr NA, Felder RA. 3D cell culture opens new dimensions in cell-based assays. Drug Discov Today 2009;14(1–2):102–7.
[70] Bicer M, Cottrell GS, Widera D. Impact of 3D cell culture on bone regeneration potential of mesenchymal stromal cells. Stem Cell Res Ther 2021;12(1):31.
[71] Elashry MI, Baulig N, Wagner A-S, Klymiuk MC, Kruppke B, Hanke T, Wenisch S, Arnhold S. Combined macromolecule biomaterials together with fluid shear stress

promote the osteogenic differentiation capacity of equine adipose-derived mesenchymal stem cells. Stem Cell Res Ther 2021;12(1):116.

[72] Li M, Ma J, Gao Y, Dong M, Zheng Z, Li Y, Tan R, She Z, Yang L. Epithelial differentiation of human adipose-derived stem cells (hASCs) undergoing three-dimensional (3D) cultivation with collagen sponge scaffold (CSS) via an indirect co-culture strategy. Stem Cell Res Ther 2020;11(1):141.

[73] Bovolenta P, Wandosell F, Nieto-Sampedro M. CNS glial scar tissue: a source of molecules which inhibit central neurite outgrowth. In: Neuronal-astrocytic interactions - implications for normal and pathological CNS function; 1992. p. 367–79 [Chapter 31].

[74] Kushchayev SV, Giers MB, Hom Eng D, Martirosyan NL, Eschbacher JM, Mortazavi MM, Theodore N, Panitch A, Preul MC. Hyaluronic acid scaffold has a neuroprotective effect in hemisection spinal cord injury. J Neurosurg Spine 2016;25(1):114–24.

[75] Fouad K, Schnell L, Bunge MB, Schwab ME, Liebscher T, Pearse DD. Combining Schwann cell bridges and olfactory-ensheathing glia grafts with chondroitinase promotes locomotor recovery after complete transection of the spinal cord. J Neurosci 2005;25(5):1169–78.

[76] Fallica B, Maffei JS, Villa S, Makin G, Zaman M. Alteration of cellular behavior and response to PI3K pathway inhibition by culture in 3D collagen gels. PLoS One 2012;7(10), e48024.

[77] Madaghiele M, Sannino A, Yannas IV, Spector M. Collagen-based matrices with axially oriented pores. J Biomed Mater Res A 2008;85(3):757–67.

[78] Yoo J, Kim HS, Hwang DY. Stem cells as promising therapeutic options for neurological disorders. J Cell Biochem 2013;114(4):743–53.

[79] Lv B, Zhang X, Yuan J, Chen Y, Ding H, Cao X, Huang A. Biomaterial-supported MSC transplantation enhances cell-cell communication for spinal cord injury. Stem Cell Res Ther 2021;12(1):36.

[80] Xiao Z, Tang F, Zhao Y, Han G, Yin N, Li X, Chen B, Han S, Jiang X, Yun C, Zhao C, Cheng S, Zhang S, Dai J. Significant improvement of acute complete spinal cord injury patients diagnosed by a combined criteria implanted with NeuroRegen scaffolds and mesenchymal stem cells. Cell Transplant 2018;27(6):907–15.

[81] Maahs DM, West NA, Lawrence JM, Mayer-Davis EJ. Epidemiology of type 1 diabetes. Endocrinol Metab Clin N Am 2010;39(3):481–97.

[82] Ling C. Epigenetic regulation of insulin action and secretion - role in the pathogenesis of type 2 diabetes. J Intern Med 2020;288(2):158–67.

[83] de Witte SF, Franquesa M, Baan CC, Hoogduijn MJ. Toward development of iMesenchymal stem cells for immunomodulatory therapy. Front Immunol 2015;6:648.

[84] Krawiec JT, Vorp DA. Adult stem cell-based tissue engineered blood vessels: a review. Biomaterials 2012;33(12):3388–400.

[85] Joggerst SJ, Hatzopoulos AK. Stem cell therapy for cardiac repair: benefits and barriers. Expert Rev Mol Med 2009;11, e20.

[86] Kholmukhamedov A, Logdon C, Hu J, McKinney RA, Spinale FG, Lemasters JJ, Mukherjee R. Cyclosporin A in left ventricular remodeling after myocardial infarction. Am J Physiol Heart Circ Physiol 2014;306(1):H53–9.

[87] Sanganalmath SK, Bolli R. Cell therapy for heart failure: a comprehensive overview of experimental and clinical studies, current challenges, and future directions. Circ Res 2013;113(6):810–34.

[88] Liang X, Ding Y, Zhang Y, Tse HF, Lian Q. Paracrine mechanisms of mesenchymal stem cell-based therapy: current status and perspectives. Cell Transplant 2014;23(9):1045–59.

[89] Ozpur MA, Guneren E, Canter HI, Karaaltin MV, Ovali E, Yogun FN, Baygol EG, Kaplan S. Generation of skin tissue using adipose tissue-derived stem cells. Plast Reconstr Surg 2016;137(1):134–43.

[90] Starokozhko V, Hemmingsen M, Larsen L, Mohanty S, Merema M, Pimentel RC, Wolff A, Emneus J, Aspegren A, Groothuis G, Dufva M. Differentiation of human-induced pluripotent stem cell under flow conditions to mature hepatocytes for liver tissue engineering. J Tissue Eng Regen Med 2018;12(5):1273–84.
[91] Jain E, Damania A, Kumar A. Biomaterials for liver tissue engineering. Hepatol Int 2014;8(2):185–97.
[92] Vasanthan KS, Subramanian A, Krishnan UM, Sethuraman S. Role of biomaterials, therapeutic molecules and cells for hepatic tissue engineering. Biotechnol Adv 2012;30(3):742–52.
[93] Li J, Tao R, Wu W, Cao H, Xin J, Li J, Guo J, Jiang L, Gao C, Demetriou AA, Farkas DL, Li L. 3D PLGA scaffolds improve differentiation and function of bone marrow mesenchymal stem cell-derived hepatocytes. Stem Cells Dev 2010;19(9):1427–36.
[94] Yagi H, Fukumitsu K, Fukuda K, Kitago M, Shinoda M, Obara H, Itano O, Kawachi S, Tanabe M, Coudriet GM, Piganelli JD, Gilbert TW, Soto-Gutierrez A, Kitagawa Y. Human-scale whole-organ bioengineering for liver transplantation: a regenerative medicine approach. Cell Transplant 2013;22(2):231–42.

CHAPTER 5

Mesenchymal stem cells: A promising weapon for cancer therapy

Xinrun Zhou[a,*], Wen Zhang[a,*], Yue Liu[a,b], Leisheng Zhang[a,c,d,e,f,g], and Zongjin Li[a,b,g]
[a]School of Medicine, Nankai University, Tianjin, China
[b]The Key Laboratory of Bioactive Materials, Ministry of Education, College of Life Sciences, Nankai University, Tianjin, China
[c]Key Laboratory of Molecular Diagnostics and Precision Medicine for Surgical Oncology in Gansu Province & NHC Key Laboratory of Diagnosis and Therapy of Gastrointestinal Tumor, Gansu Provincial Hospital, Lanzhou, China
[d]Key Laboratory of Radiation Technology and Biophysics, Hefei Institute of Physical Science, Chinese Academy of Sciences, Hefei, China
[e]National Engineering Technology Center for Cell Products & Jiangxi Research Center of Stem Cell Engineering, Jiangxi Health-Biotech Stem Cell Technology Co., Ltd., Shangrao, China
[f]Institute of Stem Cells, Health-Biotech (Tianjin) Stem Cell Research Institute Co., Ltd., Tianjin, China
[g]Tianjin Key Laboratory of Human Development and Reproductive Regulation, Tianjin Central Hospital of Gynecology Obstetrics, Nankai University Affiliated Hospital of Obstetrics and Gynecology, Tianjin, China

5.1 Introduction

In the early 1970s, a kind of nonhematopoietic fibroblast-like cell in the bone marrow characterized by the ability to differentiate into osteoblasts was identified by Friedenstein [1,2]. These multipotent cells were named mesenchymal stem cells (MSCs) due to their multipotency [3]. MSCs can be extracted from various tissues, including bone marrow, adipose tissue, peripheral blood, umbilical cord blood [4], Wharton's jelly [5], amniotic fluid, heart [6], and brain [7], indicating that these cells are involved in tissue repair or regeneration throughout life.

MSCs mainly exist in bone marrow, with low abundance, −0.001% to 0.01%, of the total nucleated cells. Under proper conditions, they can differentiate into determined cell types, such as chondrocytes, adipocytes, and osteocytes [8]. In addition to mesenchymal differentiation potential, MSCs also give rise to endodermal and ectodermal lineages, epithelium-like cells [9], neurons, and astrocytes [10]. Clonal studies and morphological observation of MSCs have indicated heterogeneity in both the function and

*These authors contribute equally to present work.

https://doi.org/10.1016/B978-0-323-95346-7.00001-4

composition of the cell population. For example, MSCs usually contain several cell subsets, including spindle-form fibroblast-like cells, flattened cells, or round cells [9,10].

Because of the heterogeneous nature of MSCs, their function and characterization are more related to specific methods of isolation. Additionally, they do not express unique cell surface markers that are warranted to identify MSCs within bone marrow, but rather a cluster of surface markers, such as CD29, CD44, CD73, CD90, CD105, STRO-1, SSEA-1, and SSEA-4, while CD14, CD19, CD34, CD45, CD11b, CD80, CD86, and HLA-DR [11] are hardly expressed. Moreover, MSCs are mature stem cells with the abilities of self-renewal, multilineage differentiation, and immunomodulation, making them an excellent therapeutic tool for a variety of diseases, such as osteogenic and cartilage repair, wound repair, lung injury, kidney disease, and autoimmune disease [12–16]. In summary, MSCs are pluripotent adult cells with extensive sources, inherent heterogeneity, and therapeutic capacities, and are considered a promising weapon against numerous diseases.

Cancer cells reveal multifacted characteristics, including sustained proliferation, activated invasion and metastasis, elusive evolution direction, intratumor cell heterogeneity and drug resistance, which thus results in cancer and causes more than 10 million mortalities annually worldwide [17–19]. Current treatment methods include surgical management, radiation, and chemotherapy. However, painful side effects, the possibility of recurrence, and toxicity to healthy tissues are almost inevitable [19]. As a consequence, new approaches with higher safety and efficacy are urgently needed.

According to previous studies, MSCs participate in multiple steps of tumor development, including angiogenesis, metastasis, epithelial-mesenchymal transition (EMT), and immunosuppression [20]. Importantly, there is an inherent cancer or inflammation tropism of MSCs, enabling them to precisely deliver targeted drugs to tumor regions [21]. Additionally, given that MSCs naturally regulate immune processes and exert antiinflammatory functions, they have been considered as opening a new avenue for cancer treatments.

In this review, we describe the therapeutic potential of MSCs due to their cancer tropism and anticancer effects, discuss their dual characteristics in cancer, and summarize current MSC modification strategies for cancer treatments. Finally, we focus on the specific applications of modified MSCs in cancer therapy and discuss the inspiring direction in this field for future attention.

5.2 The potential of MSCs for cancer therapy

5.2.1 MSC homing toward cancer or inflamed tissues

As previously reviewed, MSCs were shown to exhibit strong tropism to tumor or inflamed tissues, leading to underlying anticancer therapy applications. A wide number of cytokines, chemokines, and other functional molecules have been illustrated to mediate this process. First, several cytokines, such as IL-6, IL-1β, TNF-α, and IFN-γ, participate in the step of circulating MSCs adhering and traversing the vascular endothelial layer. Second, chemokines activate the ligand-receptor axis in cells, enabling the migration of MSCs toward targeted sites [22]. Stromal cell-derived factor-1 (SDF1), also known as C-X-C motif chemokine ligand 12 (CXCL12), is a common triggering factor in injured tissues, while C-X-C chemokine receptor type 4 (CXCR4) can be highly expressed by MSCs, binding to SDF1, which contributes to cell migration [23,24].

In addition, tumor cells highly express C-X-C motif chemokine ligand 16 (CXCL16) [25], a ligand of CXCR6. Currently, the SDF-1/CXCR4 and CXCL16/CXCR6 axes play the most important role in MSC homing capacities (Fig. 1). There are other chemokine receptors, such as CCR1, CCR4, CCR6, CCR7, CXCR4, CXCR6, and CX3CR1, on MSCs [26], which are also involved in this process. Third, adhesion molecules and matrix metalloproteinases, such as CD49d, CD44, CD54, CD106, VCAM-1, ICAM-1 [27], MMP-1, MMP-2, and MMP-9 [28], are related to the mobilization of MSCs.

5.2.2 Anticancer ability of MSCs

Once MSCs arrive at tumor sites, a wide variety of growth factors and cytokines are released to affect tumor development via direct or indirect mechanisms. For instance, MSCs were proven to inhibit the growth and metastasis of tumors through cell survival, cell cycle, angiogenesis, and immunomodulation via multiple paracrine factors, which play a leading role in tumor regulation [29]. Previous investigations have revealed that MSCs secrete cytotoxic agents into cancer tissues [30]. For instance, MSCs can secrete tumor necrosis factor-related apoptosis-inducing ligand (TRAIL), which was proven to perform tumor suppression functions in different types of tumors [31,32]. Akimoto et al. found that apoptosis of primary glioblastoma multiforme (GBM) could be promoted via TRAIL released by UCB-MSCs [33]. Recently, more advanced therapies relying on the

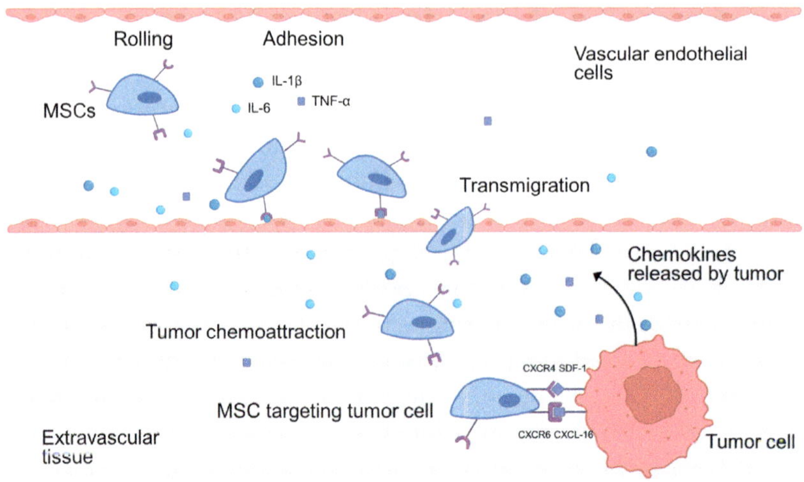

Fig. 1 The whole process of MSC homing to tumor site. A variety of cytokines and chemokines released by tumor cells are diffused in the tumor microenvironment, attracting the movement of MSCs in blood vessels. The adhesion to vascular endothelial cells and transmigration steps are introduced by chemokines such as IL-6, IL-1β, and TNF-α. After entering the tumor microenvironment, MSCs effectively target and stably bind to tumor cells via some ligand-receptor axes, such as the SDF-1/CXCR4 and CXCL16/CXCR6 axes.

combination of MSC-delivered TRAIL and paclitaxel (PTX)-based chemotherapy have emerged, which may enhance the treatment effectiveness and diminish underlying drug resistance [34].

Additionally, MSCs are able to promote apoptosis and suppress proliferation of tumor cells via several signaling pathways. Bone marrow mesenchymal stem cells (hBMSCs) were discovered to downregulate the PI3K/AKT signaling pathway, which can inhibit proliferation and accelerate apoptosis of human glioma U251 cells [35]. Recently, a report illustrated that exosomes derived from bone marrow mesenchymal stem cells could inhibit the progression of pancreatic cancer (PC) through the Wnt8/β-catenin pathway [36]. Moreover, extracellular vesicles derived from Wharton's Jelly MSC (WJ-EV) displayed the ability to regulate the HIF1α signaling pathway via miRNA-125b, weakening the tumorigenesis and metastatic ability of triple-negative breast cancer (TNBC) [37]. In another study, MSC-derived exosomes were demonstrated to successfully suppress the proliferation, migration, and invasion of hepatocellular carcinoma stem cells, which was chiefly associated with the DUSP1/ERK axis [38].

Aside from their capacity to inhibit tumor cell proliferation, studies have also shown that MSCs can diminish tumor angiogenesis, which is mediated by various signaling pathway axes. For instance, bone marrow MSCs (BM-MSCs) were documented to notably alleviate tumor neovascularization and progression in a ΔGli36 glioma xenograft mouse model. Ho et al. illustrated that downregulation of the PDGF/PDGFR axis may play a primary role in this process. Additionally, several cytokines in the tumor environment, such as IL-1β, phosphorylated Akt, and cathepsin B proteins, diminished after BM-MSC treatment [39]. Another study reported that extracellular vehicles (EVs) and conditioned medium (CM) from young donor-derived BM-MSCs showed great antiangiogenic effects in various hematologic neoplasia. In particular, the key point of high therapeutic efficacy is the age of the MSC donor, not the type of hematologic tumor [40].

Furthermore, cell-to-cell contact was also revealed to inhibit tumor growth. A report indicated that inhibitory effects on tumor development could be observed when glioblastoma-derived cancer stem cells (CSCs) and umbilical cord (UC)-MSCs were co-cultured in vitro [41]. Moreover, Guiho et al. reported that MSCs expressing TNF-related apoptosis-inducing ligand (MSC-TRAIL) showed strong apoptosis-inducing effects on Ewing sarcoma (EWS) cell lines via direct cell-to-cell contact, which was associated with a high level of death receptor (DR) clustering on the surface of EWS cells [42].

In summary, the anticancer mechanisms of MSCs can be summarized as follows: (1) delivery of cytotoxic agents, (2) regulation of cell proliferation-related signaling pathways, (3) inhibition of tumor angiogenesis, and (4) induction of apoptosis through cell-to-cell contact.

However, numerous reports have revealed that MSCs can also stimulate the growth and metastasis of tumors. For instance, MSCs can enable tumor angiogenesis [43], develop into cancer-associated fibroblasts (CAFs) in the tumor microenvironment [44], induce immune escape from the immune system of the body and accelerate tumor progression [45,46], or promote the progression of tumors through cell-to-cell contact.

Therefore, it is clear that the effects of MSCs on tumors are bidirectional. Reports have documented that the interaction between MSCs and tumors may be tightly associated with the source of MSCs, culture conditions, animal models, type of tumor, modulation of TME, and dose and time of drug treatment [47,48]. Therefore, more in vitro and in vivo investigations are urgently needed to clarify the anticancer ability of MSCs and further mechanisms. In particular, improving the precision of targeted anticancer therapies and alleviating side effects is the key for the next step.

5.3 The advantages and limitations of MSCs in cancer therapy

Compared with traditional treatment methods, MSCs exhibit several outstanding features that make them a novel choice for cell-based cancer therapy. Meanwhile, given that they play bidirectional roles in cancer, there must be a few avoidable drawbacks. In this section, we focus on the advantages and weaknesses of MSCs for cancer treatments.

As previously reviewed, the applications of MSCs in cancer therapy chiefly depend on their strong tumor tropism and natural antitumor activity [49]. In the antitumorigenic state, MSCs have been shown to induce anticancer immunity, suppress vascular regeneration, and accelerate cancer cell apoptosis [50,51]. In addition, an increasing number of studies have focused on MSC-derived extracellular vehicles (EVs), which play a key role in delivering functional molecules (cytokines, chemokines, proteins, RNAs, and lipids) in the interaction between MSCs and targeted cells [52]. MSC-derived EVs have been documented to maintain the bioactivity and tumor tropisms of MSCs and reduce the side effects of direct contact between MSCs and target cells. Moreover, unlike naturally derived MSCs, artificially modified MSCs (genetically or nongenetically) can transport specific proteins, RNAs, suicide genes, lysozyme virus, or other therapeutic molecules [53,54]. In conclusion, these features have effectively enhanced tumor specificity and reduced systemic toxicity, making MSCs a promising cell-therapy tool for cancer treatments.

However, the imperfect technology for isolation and purification of MSCs and their inherent shortages also limit their usage in cancer treatments. MSCs usually consist of multiple cell subtypes with different biological characteristics. When homing to the tumor microenvironment, the heterogeneity of MSCs may cause unknown or even negative effects for cancer therapies [20]. Furthermore, MSCs have been considered a "double-edged sword" for tumors. In addition to antitumor abilities, they can also become pro-tumorigenic under specific conditions [55]. Pro-tumor MSC phenotypes promote tumor neovascularization, facilitate tumor metastasis, and inhibit tumor cell apoptosis. Other drawbacks include off-target issues caused by inadequate targeting capability [56], side effects of residual MSCs after treatment [57], or drug resistance related to genetic modifications [53].

5.4 Modification strategies to enhance MSC therapy

Although MSCs have shown robust capacities to kill tumor cells, there are still some considerable problems that need to be urgently solved. On this basis, certain modifications of MSCs may be effective in improving their tumor targeting abilities and antitumor activities, and to reduce uncontrollability. In this part, we emphasize the gene modification strategies and briefly describe the cell membrane modification strategies.

5.4.1 Genetic modifications

Currently, genetic modifications are mature and widely used in practice. These strategies are mainly divided into two categories: viral transduction and nonviral strategies (Fig. 2).

Viral vectors

Using viral vectors to transfer target genes is based on their ability to infect cells. The most widely used viruses are retroviruses, lentiviruses, and

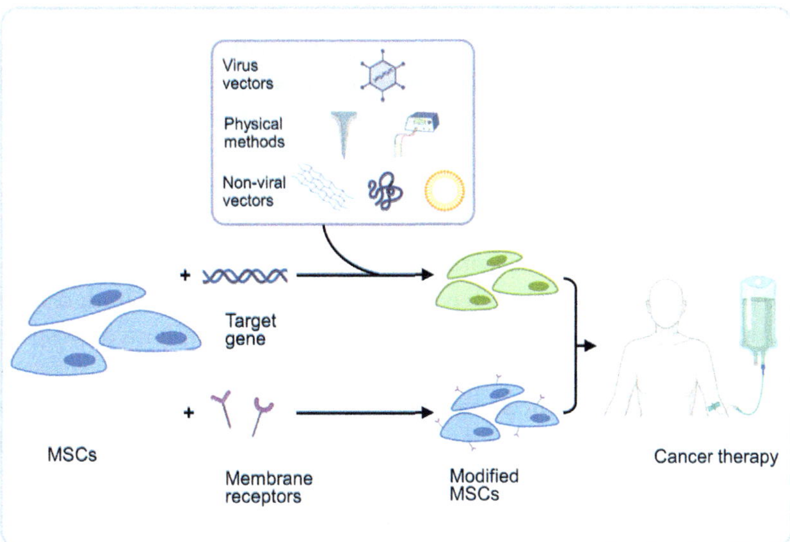

Fig. 2 The gene/membrane modification strategies of MSCs for cancer treatments.

adenoviruses, which can insert genes into the genome of the host cell or retain genes in the cytoplasm directly. Viral vectors have high transduction efficiency and stable gene expression, but they also have many side effects and significant drawbacks. Viruses can directly trigger an immune response or cause permanent heritable changes, leading to limited DNA loading and difficulty in mass production [58].

Retroviruses

Retroviruses have two identical sense single-stranded RNAs. After transfer, the virus uses reverse transcriptase to produce DNA and integrate it into the host DNA. Subsequently, gene expression occurs normally [59]. The ability to integrate genetic genes into the host genome gives it the advantage of longstanding stable expression. Retroviral vectors can transfect continuously dividing cells efficiently, while stable and permanent cells are less susceptible to infection. In recent years, retroviruses have been found to have an obvious drawback, which is that they can activate oncogenes [60].

The first clinical trial of gene therapy for X-linked severe combined immunodeficiency disease was conducted in 1999 using CD34-positive hematopoietic stem cells, which were transfected by MLV retroviral vectors to treat SCID [61]. Similar clinical trials were conducted in subsequent decades. Overall, T-cell immunity could approximately return to normal levels. However, during follow-up, some patients suffered from T-cell leukemia, and some of them even died of refractory leukemia. This may be related to the integration of retroviral vectors within or near oncogenes and activating transcription [62]. Therefore, how to improve the insertion specificity still needs to be explored.

Lentiviruses

Lentiviruses, as one family of retroviruses, also integrate target genes into the host genome. However, lentiviruses are superior vectors compared to other retroviruses [63]. Lentiviruses have been shown to transduce both dividing and dividing cells and to have effects on neurons, cardiomyocytes, etc., allowing them to be more widely used [64].

In a study in 2011, genetic modification of MSCs by various types of lentiviruses was evaluated in terms of efficiency and persistence. The results showed that the second-generation lentiviral vector rHIV-pWPT-EF1-α-GFP-W was the most efficient and had the most persistent gene expression. Overall, all data indicate that lentiviral vector transduction can promote therapeutic gene expression and provide an effective tool for in vitro MSC modification [65].

Likewise, the most common disadvantage of lentiviruses is nonspecificity, which causes mutagenesis, but this is less likely to cause oncogene activation than are retroviruses [66].

Adenoviruses

Adenoviruses have no envelope and contain a linear double-stranded genome of approximately 36 kb in a dodecahedral form [67]. Adenoviruses cannot integrate genes into the host genome, so there is no risk of insertional mutagenesis, and they have a high loading capacity. Adenoviruses can infect both resting and dividing cells and replicate in the nucleus [68].

It has been demonstrated that right ventricular hypertrophy and pulmonary vascular remodeling were significantly improved in mice injected with bone marrow-derived mesenchymal stem cells transfected with the recombinant adenoviral vector Ad-let-7a, which shows a protective effect of MSC-let-7a against PAH. However, excess reactive oxygen species (ROS) increase apoptosis of BM-MSCs, resulting in poor survival and transplantation efficiency. Therefore, it is important to improve the ability of BM-MSCs to eliminate ROS [69]. In recent years, some researchers have constructed adenoviruses expressing a short hairpin (sh) RNA directed against Mst1 (Ad-sh-Mst1) and harboring control vectors for Ad-sh-Mst1 (Ad-LacZ). These were then transfected with MSCs. The results showed that the inhibition of Mst1 reduced ROS production and improved MSC survival and transplantation efficiency [70].

One of the main drawbacks of adenoviruses is hepatotoxicity. The mechanism is the accumulation of activated immune cells, which may include Kupffer cells (KCs), neutrophils, natural killer (NK) cells, and γδ T cells, and the secretion of excessive cytokines and chemokines in the liver, ultimately leading to inflammatory damage [71]. Although adenoviruses are widely used, their immunogenicity still limits their application. Therefore, it is important to construct low-immunogenicity viral vectors [72].

Nonviral strategies

Nonviral modification strategies mainly include physical methods and nonviral vectors. In this review, we will introduce the most widely used ones.

Physical methods

The main physical gene delivery techniques are microscopic injection, electroporation, acoustic perforation, laser perforation, sonoporation, and so on. These strategies can circumvent the problems of immunogenicity and cytotoxicity, but at the same time, the cells suffer physical damage, which can lead to apoptosis. The principles of some of them are simple, but they are difficult to implement.

Microinjection This method involves directly injecting nucleic acids into the nucleus under the microscope. It is relatively simple, economical, effective, and nontoxic, and can transfer large sizes of DNA [73]. However, this method, commonly using a single microinjection needle, is inefficient because the number of target cells for genetic modifications is large. In addition, microinjection has been mostly applied to rat oocytes due to the relatively large size of oocytes, whereas other smaller somatic cells cannot accommodate conventional microinjection needles [74].

Electroporation Electroporation is currently the most widely used physical delivery method and is considered the most effective. In this way, a population of cells is placed in a transient electric field that can alter the permeability of the cell membrane to allow nucleic acids to enter the cytoplasm [75]. Studies have demonstrated the feasibility and safety of electroporation for the delivery of genes to MSCs. In one study, PD-MSCs overexpressing PRL-1 were constructed via electroporation, and a lentivirus vector was used as a control group. The results showed that the transfection efficiency by electroporation was equal to that of lentivirus. The expression of mRNA was increased, especially significantly before the eighth generation. The dose of electroporation buffer solution had no effect on cell viability or differentiation potential [76]. However, it has also been shown that transient electric fields have some effect on some molecules, including lipid membranes, proteins, and DNA [75].

Sonoporation This method is similar to electroporation, in that it uses ultrasound to create small pores in the cell membrane, thereby allowing nucleic acids to enter the cell [77]. Sonoporation is a passive diffusion, and the efficiency of transmission depends on the ultrasound frequency and intensity [78]. Sonoporation is noninvasive. It can be accomplished by local ultrasound irradiation without the need for surgery [78]. Currently, sonoporation is mainly used for drug delivery and gene therapy [79].

However, similarly to electroporation, sonoporation disrupts the integrity of the cell membrane and, therefore, has the potential to cause apoptosis, and the efficiency of gene transfer is low [78]. This may be associated with delayed DNA synthesis and prolongation of the cell cycle [80].

Nonviral vectors

Nonviral vectors can circumvent some of the drawbacks of viral vectors. They are less immunogenic, less mutagenic, and easier to generate, but their gene transfer efficiency and specificity are lower. The duration of gene expression may be shorter, and their safety profile must be considered. Currently, the most commonly investigated nonviral vectors are polymers, lipids, peptides, inorganic particles, or hybrid systems [81]. Cationic vectors are capable of maximizing the capture and binding of target genes and can form stable complex compounds without disrupting their integrity [82]. Cationic lipids and cationic polymers are the most widely used vectors [73].

Cationic lipids Most lipids contain positively charged head groups that can bind to the anionic phosphate groups of nucleic acids to form compounds by electrostatic action. Lipids have the advantages of being biodegradable, having low immunogenicity, and absorbing hydrophobic substances [83]. Cationic lipids consist of three components: the head group, linker, and tails. Among these, the polar head group is dominant in binding to the anionic NAs. The linker, located between the hydrophilic polar head group and the nonpolar tail group, has an impact on stability, biodegradability, cytotoxicity, and transfection efficiency [84]. The tail portion is the hydrophobic region and is associated with the mobility, overall stability, and cytotoxicity of the lipid complex formed [85]. A single cationic lipid can deliver nucleic acids but is usually bound to a co-lipid to increase effectiveness. Additionally, the presence of a co-lipid has been shown to reduce the cytotoxicity caused by the nonspecific binding of cationic lipids to serum proteins [86].

Cationic polymorphs The charged amine groups of cationic polymorphs can be protonated and then can bind to the negative charges of DNA to form polymer-DNA compounds [87]. Poly-ethylenimine (PEI) is the first polycationic polymer [88]. However, its lack of specificity and transfection properties limits its application, and it is a nonbiodegradable polymer that is prone to cytotoxicity [58]. In addition to PEI, amine-terminated PAMAM is also a dendrimer. In fact, PAMAM is one of the most commonly used dendrimers in biological applications. It was the first to be used for gene delivery, but cytotoxicity (related to the chemical nature of the surface amine group) limits its application. To address this issue, in one study, researchers modified PAMAM with hyaluronic acid (HA). Next, they used multiple modalities to assess the cytotoxicity and transfection efficiency. The results revealed that the cytotoxicity of the polymers was reduced after HA modification [89].

Peptides Peptides are short chains of $\leq$50 amino acid residues [90]. Genetic material interacts with peptides by coupling or electrostatic forces to form peptide-nucleic acid compounds [58]. These are stable and uncharged molecules, which makes them resistant to degradation by nucleases and less sensitive to acidic, alkaline, and high temperature environments. Because of their stability, low cytotoxicity, and immunogenicity, peptides are widely used as nonviral gene carriers [90]. The most widely used peptide is PLL, which has a high side chain charge density and binds to DNA at these sites. It has been shown that PLL-DNA compounds are easily absorbed by endocytosis [91].

In recent years, a number of functional peptides have also been developed to overcome physiological barriers to improve delivery efficiency. For example, CPPs are short peptides that have the potential to penetrate biological membranes and deliver a variety of cargoes into the cell. Many CPPs are derived from protein sequences such as TAT, M918, and VP22, but currently, TAT is the most popular [92]. Although CPPs can penetrate cells well, they are retained in endosomes, leading to degradation. To address this, modifications can be conducted. Dioleoyl phosphatidylethanolamine (DOPE), as an auxiliary lipid, is applicable [90].

5.4.2 Cell membrane modification

Compared to genetic modifications, cell membrane modifications are not very widely used. The membrane surface receptors of MSCs include cytokine receptors, chemokine receptors, growth factor receptors, cell matrix receptors, and cell-to-cell interaction receptors. Currently, MSCs are widely applied in cancer therapy due to their ability to migrate to tumor cells, which enables them to be used as a vehicle to translate therapeutic drugs, genes, or oncolytic viruses. The binding of cytokines to cytokine receptors can mediate adhesion and crossing of vascular endothelial cells, and the binding of chemokines to chemokine receptors can activate chemotaxis in MSCs to cause migration toward tumor cells [93]. To improve the efficiency and accuracy of MSC migration, appropriate modifications of the cell membrane are necessary. In a study in 2021, researchers linked "targeting peptide" (TP), catalyzed by Tgase, to proteins naturally expressed in the MSC cell membrane via a covalent method. The results showed that the cell membrane modification technique facilitated the targeting of MSCs to inflammatory tissue and enhanced adhesion [94].

5.5 Modified MSCs in cancer therapeutics

5.5.1 MSCs as anticancer agent delivery tools

Regarding the biological properties of MSCs, it has been mentioned that MSCs have the ability to migrate toward cancer cells and inflammatory tissues directly. MSCs have been used as carriers of anticancer drugs to deliver chemical agents for tumor therapy due to their distinct tropism [93]. In addition, the low immunogenicity of MSCs promotes their application in tumor therapy. The cargoes carried by MSCs can be divided into two main categories. One kind is free chemical drugs, and the other kind is nanoparticles (NPs) in which anticancer drugs are encapsulated [95].

Regarding the delivery of free drugs by MSCs, we take paclitaxel (PTX) as an example. PTX, as one of the most common chemotherapeutic agents, has been applied in tumor treatments for many years. PTX is taken up into MSCs by co-culturing, and then MSCs migrate to the tumor tissue and deliver PTX to the cancer cells via secretory vesicles. A study also demonstrated that secretory vesicles containing PTX had visible antitumor properties in vitro, which laid a theoretical foundation for perceiving MSC-derived extracellular vesicles as a tool for drug delivery [96].

The application of MSCs is limited by the short retention time of the drugs after being loaded by MSCs. To overcome this drawback, it has been proposed that NPs can be utilized to enclose drugs, which are then internalized by MSCs. In this way, sustained-release chemotherapeutic drugs are available. In short, the use of NPs that can sustainably release drugs, which contributes to extending the retention time of the loaded drugs in MSCs [97]. The following is an example of paclitaxel-poly (lactic-*co*-glycolic acid) nanoparticles. In this study, researchers obtained PTX-PLGA NP-loaded mesenchymal stem cells by incubating them with PTX-PLGA NPs. Next, MSCs were injected into a glioma rat model to investigate their antitumor ability. The results showed that the optimal dose of MSC-PTX-PLGA NPs (1 pg/cell PTX) had only a slight effect on the migratory capacity, cell cycle, and differentiation potential of MSCs. The retention time of the drugs was prolonged in this modality compared to the treatment modality of MSCs loaded with free PTX. Based on experimental statistics, the PTX-PLGA NP-loaded MSCs were more effective in inhibiting glioma growth. Therefore, internalization of chemotherapeutic drug-loaded NPs into MSCs is a promising strategy for tumor-targeted therapy [98].

However, as mentioned above, the tumor-targeting capacity of MSCs is limited after cargo loading, so some appropriate modifications of MSCs are necessary. Various modifications have been introduced in the "MSC modification strategy." Nevertheless, internalizing the cargoes may not only affect the biological activity of MSCs but also lead to cytotoxicity. Thus, a novel attempt to add drugs to the cell membrane surface of MSCs by molecular modification has emerged [21]. In one study, researchers attempted to modify the surface of MSCs with doxorubicin-loaded liposomes (DOX-Lips) through the avidin-biotin complex method (ABC method). DOX-Lip-modified C3H10T1/2 cells (a mouse MSC line), with approximately 21.5 pg of DOX per cell, had little effect on the tumor migration capacity of C3H10T1/2 cells. The study also showed that DOX-Lip-modified C3H10T1/2 cells distinctly inhibited the proliferation of mouse colon adenocarcinoma colon26/fluc cells compared with DOX-Lips alone. In addition, DOX-Lip-modified C3H10T1/2 cells inhibited tumor growth in a mouse model of subcutaneous tumors and lung metastasis tumors [99].

5.5.2 MSCs release therapeutic proteins

Currently, protein-based therapy is widely used to treat various diseases, such as tumors, immune disorders, infections, and other diseases. According to molecular types, therapeutic proteins can be classified as monoclonal antibodies, Fc fusion proteins, enzymes, growth factors, hormones, interferons, interleukins, etc. [100]. MSCs can be used as delivery vehicles to transport them to target cells.

TRAIL is a protein produced by most normal tissues when cancer cells are generated. TRAIL binds to the death receptors TRAIL-R1 (DR4) and TRAIL-R2 (DR5), promoting the formation of the death-inducing signaling complex (DISC), which ultimately leads to cell apoptosis [101]. However, there are still some problems with using TRAIL alone. Notably, a considerable proportion of tumor cells are TRAIL resistant. To overcome this resistance, some investigators have tried to make use of MSCs to deliver TRAIL and use TRAIL-resistant colon cancer cells as targets to explore its effect. The results showed that TRAIL-MSCs could overcome TRAIL resistance in selected colon cancer cells through direct cell-to-cell interactions. Thus, utilizing MSCs as delivery vehicles to load TRAIL is an effective clinical tool to overcome TRAIL resistance [102].

In another study, using pancreatic ductal adenocarcinoma (PDAC) as an example, MSCs were modified genetically via viral vectors to express a novel secretory TRAIL variant, which has been verified to induce apoptosis [54]. On this basis, more combined therapy has emerged to improve the treatment effect of MSCs-TRAIL. For example, some investigators have explored the effectiveness of valproic acid (VPA) in improving apoptosis induction of TRAIL. In experiments, the combined application of VPA and MSCs-TRAIL resulted in increased MSC migration to tumor sites in glioma mice compared with VPA or TRAIL-MSCs alone. In vivo experiments indicated that concurrent treatment with VPA and MSCs-TRAIL also had greater efficacy than treatment with each agent alone [103]. Another study demonstrated that, in the combination of TRAIL-MSCs and PTX, PTX was able to restore the sensitivity of pancreatic cancer to TRAIL released from MSCs, as well as augmenting the cytotoxicity against tumor cells [34]. In addition, interleukins have been revealed to have antitumor capability. In one study, MSCs were genetically modified to co-express IL-12 and IL-7, and efficacy was assessed using a glioma mouse model. The results showed a significant inhibition of tumor growth and remission of established intracranial tumors [104].

5.5.3 MSCs as oncolytic adenovirus carriers

Oncolytic viruses (i.e., viruses that selectively replicate and destroy cancer cells with limited toxicity or nontoxicity to normal cells) have emerged as a new strategy for the treatment of advanced cancer or metastatic tumors. However, targeted delivery of oncolytic viruses is the main research problem. Not all oncolytic viruses (OVs) can be injected locally, and when injected systemically, only a very small fraction can migrate and reach the target tumor site, since the immune system can clear most of them. MSCs have been proposed as delivery vehicles due to their outstanding tumor targeting property and their modulation of the tumor microenvironment [105]. Ten years ago, researchers evaluated the immune response to oncolytic adenovirus (Ad)-loaded MSCs using a semipermissive cotton rat (CR) model. The results showed that MSCs could support the replication of oncolytic adenovirus, and the inhibition of T-cell proliferation was not affected. In other words, compared with oncolytic adenovirus alone, delivery by MSCs ensures normal replication of oncolytic viruses and helps them escape the immune system [106].

Furthermore, how to support viral replication better and balance the vitality of MSCs with viral replication is a question that needs to be explored. First, we can modify the oncolytic adenovirus. Interleukin-24 (IL-24), also known as melanoma differentiation-associated gene 7 (MDA-7), a member of the IL-10 family, is a kind of tumor therapeutic gene expression product carried by viral vectors. IL-24 can selectively inhibit tumor growth. Endothelin (ES), an effective inhibitor of vascular endothelial cell proliferation, can inhibit tumor angiogenesis, induce apoptosis, and inhibit tumor growth [107–110]. In one study, researchers loaded a novel oncolytic adenovirus carrying interleukin (IL)-24 and/or endothelin into human umbilical cord blood-mesenchymal stem cells (hUCB-MSCs) for glioma treatment. Statistics have shown that this novel oncolytic adenovirus can effectively replicate and kill glioma cells without harming normal cells in vitro [111].

In addition, MSCs can be modified. In another study, MSCs were modified with adenovirus (type C) E1A/B (E1s) genes (MSCs-E1s), which are essential for adenovirus replication, and carried conditionally replicating oncolytic adenovirus or replication-deficient oncolytic adenoviruses that contain p14ARF/p53 tumor suppressor genes. The results showed that MSCs-E1s could support the replication of conditionally replicating and replication-deficient oncolytic adenovirus. Moreover, the tumor tropism and dissemination of MSCs-E1s have been improved. Replication-defective adenovirus expressing p14 and p53 or conditionally replicating oncolytic adenovirus-loaded E1-modified mesenchymal stem cells suppress tumor growth in mice, providing an effective, nontoxic therapeutic alternative for metastatic disease [112].

5.6 Conclusion and future perspectives

Mesenchymal stem cells have been exploited to treat a wide variety of diseases due to their self-renewal, multipotential differentiation, immunomodulation, and antiinflammatory abilities. Cancer is a destructive and fatal disease, but characteristics such as tumor tropisms and antitumor abilities enable MSC to be an advanced therapeutic tool against cancer. However, the application of MSC in cancer therapies has been hindered by some drawbacks, like contradictory effects of MSC toward cancer. Hence, MSCs still needs to be optimized as a new weapon for cancer treatment. Genetic or nongenetic modifications can be used to manufacture ideal therapeutic MSCs for cancer treatments. Methods of MSC modifications can be used, including genetically modification (using virus or nonvirus) and cell membrane modifications, to produce MSCs that can delivery anticancer agents, or secrete therapeutic proteins or oncolytic adenovirus to kill tumor cells.

In summary, MSCs are generally perceived as a promising approach for cancer treatments in light of their modifiable properties. However, some shortages are still urgently to be overcome, such as the heterogeneity of MSCs populations or unknown effects on irrelevant tissues after MSCs enter the body. In order to maximize the antitumor capacities, the mechanisms of interaction between MSCs and tumors need to be further clarified. In addition, more suitable cell sources, culture conditions, and administration methods of MSCs for different tumors should be explored. In this context, the existing methods of MSC modification and current usages have been described. Future research should focus on MSC-based cell-free therapy for cancer, which could more efficiently delivery therapeutic vesicles and reduce adverse effects. We hope and believe that this quickly evolving field will produce more safe and effective agents and bring light to numerous cancer patients worldwide.

Acknowledgements

This study was financially supported by the National Natural Science Foundation of China (U2004126), the Tianjin Natural Science Foundation (22JCZXJC00170, 21JCZDJC00070), Tianjin Key Medical Discipline (Specialty) Construction Project (No. TJYXZDXK-043A), Key project funded by Department of Science and Technology of Shangrao City (2022A001), and Double Thousand Plan of Jiangxi Province (2022 to LSZ).

References

[1] Friedenstein AJ, Chailakhjan RK, Lalykina KS. The development of fibroblast colonies in monolayer cultures of guinea-pig bone marrow and spleen cells. Cell Tissue Kinet 1970;3(4):393–403.
[2] Friedenstein AJ, Chailakhyan RK, Latsinik NV, Panasyuk AF, Keiliss-Borok IV. Stromal cells responsible for transferring the microenvironment of the hemopoietic tissues. Cloning in vitro and retransplantation in vivo. Transplantation 1974;17(4):331–40.
[3] Caplan AI. Mesenchymal stem cells. J Orthop Res 1991;9(5):641–50.
[4] Harris DT. Umbilical cord tissue mesenchymal stem cells: characterization and clinical applications. Curr Stem Cell Res Ther 2013;8(5):394–9.
[5] De Bruyn C, Najar M, Raicevic G, Meuleman N, Pieters K, Stamatopoulos B, et al. A rapid, simple, and reproducible method for the isolation of mesenchymal stromal cells from Wharton's jelly without enzymatic treatment. Stem Cells Dev 2011;20(3): 547–57.
[6] Pilato CA, Stadiotti I, Maione AS, Saverio V, Catto V, Tundo F, et al. Isolation and characterization of cardiac mesenchymal stromal cells from endomyocardial bioptic samples of arrhythmogenic cardiomyopathy patients. J Vis Exp 2018;132, 57263.
[7] da Silva Meirelles L, Chagastelles PC, Nardi NB. Mesenchymal stem cells reside in virtually all post-natal organs and tissues. J Cell Sci 2006;119(Pt 11):2204–13.
[8] Zha K, Tian Y, Panayi AC, Mi B, Liu G. Recent advances in enhancement strategies for osteogenic differentiation of mesenchymal stem cells in bone tissue engineering. Front Cell Dev Biol 2022;10, 824812.

[9] Spees JL, Olson SD, Ylostalo J, Lynch PJ, Smith J, Perry A, et al. Differentiation, cell fusion, and nuclear fusion during ex vivo repair of epithelium by human adult stem cells from bone marrow stroma. Proc Natl Acad Sci U S A 2003;100(5):2397–402.
[10] Phinney DG, Isakova I. Plasticity and therapeutic potential of mesenchymal stem cells in the nervous system. Curr Pharm Des 2005;11(10):1255–65.
[11] Machado Cde V, Telles PD, Nascimento IL. Immunological characteristics of mesenchymal stem cells. Rev Bras Hematol Hemoter 2013;35(1):62–7.
[12] Lwin SM, Snowden JA, Griffiths CEM. The promise and challenges of cell therapy for psoriasis. Br J Dermatol 2021;185(5):887–98.
[13] Yun CW, Lee SH. Potential and therapeutic efficacy of cell-based therapy using mesenchymal stem cells for acute/chronic kidney disease. Int J Mol Sci 2019;20(7):1619.
[14] Kaspi H, Semo J, Abramov N, Dekel C, Lindborg S, Kern R, et al. MSC-NTF (NurOwn®) exosomes: a novel therapeutic modality in the mouse LPS-induced ARDS model. Stem Cell Res Ther 2021;12(1):72.
[15] Bojanic C, To K, Hatoum A, Shea J, Seah KTM, Khan W, et al. Mesenchymal stem cell therapy in hypertrophic and keloid scars. Cell Tissue Res 2021;383(3):915–30.
[16] Fuggle NR, Cooper C, Oreffo ROC, Price AJ, Kaux JF, Maheu E, et al. Alternative and complementary therapies in osteoarthritis and cartilage repair. Aging Clin Exp Res 2020;32(4):547–60.
[17] Graham TA, Sottoriva A. Measuring cancer evolution from the genome. J Pathol 2017;241(2):183–91.
[18] Gilbertson RJ. Mapping cancer origins. Cell 2011;145(1):25–9.
[19] Zaimy MA, Saffarzadeh N, Mohammadi A, Pourghadamyari H, Izadi P, Sarli A, et al. New methods in the diagnosis of cancer and gene therapy of cancer based on nanoparticles. Cancer Gene Ther 2017;24(6):233–43.
[20] Timaner M, Tsai KK, Shaked Y. The multifaceted role of mesenchymal stem cells in cancer. Semin Cancer Biol 2020;60:225–37.
[21] Takayama Y, Kusamori K, Nishikawa M. Mesenchymal stem/stromal cells as next-generation drug delivery vehicles for cancer therapeutics. Expert Opin Drug Deliv 2021;18(11):1627–42.
[22] Stone MJ, Hayward JA, Huang C, Huma ZE, Sanchez J. Mechanisms of regulation of the chemokine-receptor network. Int J Mol Sci 2017;18(2):342.
[23] Ma J, Liu N, Yi B, Zhang X, Gao BB, Zhang Y, et al. Transplanted hUCB-MSCs migrated to the damaged area by SDF-1/CXCR4 signaling to promote functional recovery after traumatic brain injury in rats. Neurol Res 2015;37(1):50–6.
[24] Wynn RF, Hart CA, Corradi-Perini C, O'Neill L, Evans CA, Wraith JE, et al. A small proportion of mesenchymal stem cells strongly expresses functionally active CXCR4 receptor capable of promoting migration to bone marrow. Blood 2004;104(9):2643–5.
[25] Lau TT, Wang DA. Stromal cell-derived factor-1 (SDF-1): homing factor for engineered regenerative medicine. Expert Opin Biol Ther 2011;11(2):189–97.
[26] Sordi V, Malosio ML, Marchesi F, Mercalli A, Melzi R, Giordano T, et al. Bone marrow mesenchymal stem cells express a restricted set of functionally active chemokine receptors capable of promoting migration to pancreatic islets. Blood 2005;106(2):419–27.
[27] De Ugarte DA, Alfonso Z, Zuk PA, Elbarbary A, Zhu M, Ashjian P, et al. Differential expression of stem cell mobilization-associated molecules on multi-lineage cells from adipose tissue and bone marrow. Immunol Lett 2003;89(2–3):267–70.
[28] Hu YL, Fu YH, Tabata Y, Gao JQ. Mesenchymal stem cells: a promising targeted-delivery vehicle in cancer gene therapy. J Control Release 2010;147(2):154–62.
[29] Hmadcha A, Martin-Montalvo A, Gauthier BR, Soria B, Capilla-Gonzalez V. Therapeutic potential of mesenchymal stem cells for cancer therapy. Front Bioeng Biotechnol 2020;8:43.

[30] Khakoo AY, Pati S, Anderson SA, Reid W, Elshal MF, Rovira II, et al. Human mesenchymal stem cells exert potent antitumorigenic effects in a model of Kaposi's sarcoma. J Exp Med 2006;203(5):1235–47.
[31] Fakiruddin KS, Ghazalli N, Lim MN, Zakaria Z, Abdullah S. Mesenchymal stem cell expressing TRAIL as targeted therapy against sensitised tumour. Int J Mol Sci 2018;19(8).
[32] Takeda K, Hayakawa Y, Smyth MJ, Kayagaki N, Yamaguchi N, Kakuta S, et al. Involvement of tumor necrosis factor-related apoptosis-inducing ligand in surveillance of tumor metastasis by liver natural killer cells. Nat Med 2001;7(1):94–100.
[33] Akimoto K, Kimura K, Nagano M, Takano S, Salazar GT, Yamashita T, et al. Umbilical cord blood-derived mesenchymal stem cells inhibit, but adipose tissue-derived mesenchymal stem cells promote, glioblastoma multiforme proliferation. Stem Cells Dev 2013;22(9):1370–86.
[34] Rossignoli F, Spano C, Grisendi G, Foppiani EM, Golinelli G, Mastrolia I, et al. MSC-delivered soluble TRAIL and paclitaxel as novel combinatory treatment for pancreatic adenocarcinoma. Theranostics 2019;9(2):436–48.
[35] Lu L, Chen G, Yang J, Ma Z, Yang Y, Hu Y, et al. Bone marrow mesenchymal stem cells suppress growth and promote the apoptosis of glioma U251 cells through down-regulation of the PI3K/AKT signaling pathway. Biomed Pharmacother 2019;112, 108625.
[36] Yao X, Mao Y, Wu D, Zhu Y, Lu J, Huang Y, et al. Exosomal circ_0030167 derived from BM-MSCs inhibits the invasion, migration, proliferation and stemness of pancreatic cancer cells by sponging miR-338-5p and targeting the Wif1/Wnt8/β-catenin axis. Cancer Lett 2021;512:38–50.
[37] Chang YH, Vuong CK, Ngo NH, Yamashita T, Ye X, Futamura Y, et al. Extracellular vesicles derived from Wharton's Jelly mesenchymal stem cells inhibit the tumor environment via the miR-125b/HIF1α signaling pathway. Sci Rep 2022;12 (1):13550.
[38] Gu H, Yan C, Wan H, Wu L, Liu J, Zhu Z, et al. Mesenchymal stem cell-derived exosomes block malignant behaviors of hepatocellular carcinoma stem cells through a lncRNA C5orf66-AS1/microRNA-127-3p/DUSP1/ERK axis. Hum Cell 2021;34(6):1812–29.
[39] Ho IA, Toh HC, Ng WH, Teo YL, Guo CM, Hui KM, et al. Human bone marrow-derived mesenchymal stem cells suppress human glioma growth through inhibition of angiogenesis. Stem Cells 2013;31(1):146–55.
[40] Gladkova N, Umezu T, Imanishi S, Kawana C, Ohyashiki JH, Ohyashiki K. Effect of the extracellular component of bone marrow mesenchymal stromal cells from healthy donors on hematologic neoplasms and their angiogenesis. Hum Cell 2020;33 (3):599–609.
[41] Bajetto A, Pattarozzi A, Corsaro A, Barbieri F, Daga A, Bosio A, et al. Different effects of human umbilical cord mesenchymal stem cells on glioblastoma stem cells by direct cell interaction or via released soluble factors. Front Cell Neurosci 2017;11:312.
[42] Guiho R, Biteau K, Grisendi G, Taurelle J, Chatelais M, Gantier M, et al. TRAIL delivered by mesenchymal stromal/stem cells counteracts tumor development in orthotopic Ewing sarcoma models. Int J Cancer 2016;139(12):2802–11.
[43] Sun Z, Wang S, Zhao RC. The roles of mesenchymal stem cells in tumor inflammatory microenvironment. J Hematol Oncol 2014;7:14.
[44] Barcellos-de-Souza P, Comito G, Pons-Segura C, Taddei ML, Gori V, Becherucci V, et al. Mesenchymal stem cells are recruited and activated into carcinoma-associated fibroblasts by prostate cancer microenvironment-derived TGF-β1. Stem Cells 2016;34(10):2536–47.
[45] Kim JS, Kim JG, Moon MY, Jeon CY, Won HY, Kim HJ, et al. Transforming growth factor-beta1 regulates macrophage migration via RhoA. Blood 2006;108(6):1821–9.

[46] Weber F, Byrne SN, Le S, Brown DA, Breit SN, Scolyer RA, et al. Transforming growth factor-beta1 immobilises dendritic cells within skin tumours and facilitates tumour escape from the immune system. Cancer Immunol Immunother 2005;54 (9):898–906.
[47] Lan T, Luo M, Wei X. Mesenchymal stem/stromal cells in cancer therapy. J Hematol Oncol 2021;14(1):195.
[48] Zhang J, Yuan Z, Zhong W, Wei Y. Stem cell as vehicles of antibody in treatment of lymphoma: a novel and potential targeted therapy. Stem Cell Rev Rep 2021;17 (3):829–41.
[49] Zimmerlin L, Park TS, Zambidis ET, Donnenberg VS, Donnenberg AD. Mesenchymal stem cell secretome and regenerative therapy after cancer. Biochimie 2013;95 (12):2235–45.
[50] Ohlsson LB, Varas L, Kjellman C, Edvardsen K, Lindvall M. Mesenchymal progenitor cell-mediated inhibition of tumor growth in vivo and in vitro in gelatin matrix. Exp Mol Pathol 2003;75(3):248–55.
[51] Abd-Allah SH, Shalaby SM, El-Shal AS, Elkader EA, Hussein S, Emam E, et al. Effect of bone marrow-derived mesenchymal stromal cells on hepatoma. Cytotherapy 2014;16(9):1197–206.
[52] Weng Z, Zhang B, Wu C, Yu F, Han B, Li B, et al. Therapeutic roles of mesenchymal stem cell-derived extracellular vesicles in cancer. J Hematol Oncol 2021;14(1):136.
[53] Rossignoli F, Grisendi G, Spano C, Golinelli G, Recchia A, Rovesti G, et al. Inducible Caspase9-mediated suicide gene for MSC-based cancer gene therapy. Cancer Gene Ther 2019;26(1–2):11–6.
[54] Spano C, Grisendi G, Golinelli G, Rossignoli F, Prapa M, Bestagno M, et al. Soluble TRAIL armed human MSC as gene therapy for pancreatic cancer. Sci Rep 2019;9(1):1788.
[55] Betancourt AM. New cell-based therapy paradigm: induction of bone marrow-derived multipotent mesenchymal stromal cells into pro-inflammatory MSC1 and anti-inflammatory MSC2 phenotypes. Adv Biochem Eng Biotechnol 2013;130: 163–97.
[56] De Becker A, Riet IV. Homing and migration of mesenchymal stromal cells: how to improve the efficacy of cell therapy? World J Stem Cells 2016;8(3):73–87.
[57] Noble M. Can neural stem cells be used as therapeutic vehicles in the treatment of brain tumors? Nat Med 2000;6(4):369–70.
[58] Oggu GS, Sasikumar S, Reddy N, Ella KKR, Rao CM, Bokara KK. Gene delivery approaches for mesenchymal stem cell therapy: strategies to increase efficiency and specificity. Stem Cell Rev Rep 2017;13(6):725–40.
[59] Hull R, Covey SN. Genome organization and expression of reverse transcribing elements: variations and a theme. J Gen Virol 1986;67(Pt 9):1751–8.
[60] Sage EK, Thakrar RM, Janes SM. Genetically modified mesenchymal stromal cells in cancer therapy. Cytotherapy 2016;18(11):1435–45.
[61] Cavazzana-Calvo M, Hacein-Bey S, de Saint Basile G, Gross F, Yvon E, Nusbaum P, et al. Gene therapy of human severe combined immunodeficiency (SCID)-X1 disease. Science 2000;288(5466):669–72.
[62] Hacein-Bey-Abina S, Von Kalle C, Schmidt M, McCormack MP, Wulffraat N, Leboulch P, et al. LMO2-associated clonal T cell proliferation in two patients after gene therapy for SCID-X1. Science 2003;302(5644):415–9.
[63] Naldini L, Blömer U, Gallay P, Ory D, Mulligan R, Gage FH, et al. In vivo gene delivery and stable transduction of nondividing cells by a lentiviral vector. Science 1996;272(5259):263–7.
[64] Naldini L. Lentiviruses as gene transfer agents for delivery to non-dividing cells. Curr Opin Biotechnol 1998;9(5):457–63.

[65] McGinley L, McMahon J, Strappe P, Barry F, Murphy M, O'Toole D, et al. Lentiviral vector mediated modification of mesenchymal stem cells & enhanced survival in an in vitro model of ischaemia. Stem Cell Res Ther 2011;2(2):12.
[66] Montini E, Cesana D, Schmidt M, Sanvito F, Ponzoni M, Bartholomae C, et al. Hematopoietic stem cell gene transfer in a tumor-prone mouse model uncovers low genotoxicity of lentiviral vector integration. Nat Biotechnol 2006;24(6):687–96.
[67] Hoeben RC, Uil TG. Adenovirus DNA replication. Cold Spring Harb Perspect Biol 2013;5(3), a013003.
[68] Athanasopoulos T, Munye MM, Yáñez-Muñoz RJ. Nonintegrating gene therapy vectors. Hematol Oncol Clin North Am 2017;31(5):753–70.
[69] Cheng G, Wang X, Li Y, He L. Let-7a-transfected mesenchymal stem cells ameliorate monocrotaline-induced pulmonary hypertension by suppressing pulmonary artery smooth muscle cell growth through STAT3-BMPR2 signaling. Stem Cell Res Ther 2017;8(1):34.
[70] Zhang Q, Cheng X, Zhang H, Zhang T, Wang Z, Zhang W, et al. Dissecting molecular mechanisms underlying H(2)O(2)-induced apoptosis of mouse bone marrow mesenchymal stem cell: role of Mst1 inhibition. Stem Cell Res Ther 2020;11(1):526.
[71] Wang Z, Zhang X. Adenovirus vector-attributed hepatotoxicity blocks clinical application in gene therapy. Cytotherapy 2021;23(12):1045–52.
[72] Sakurai H, Kawabata K, Sakurai F, Nakagawa S, Mizuguchi H. Innate immune response induced by gene delivery vectors. Int J Pharm 2008;354(1–2):9–15.
[73] Wang W, Xu X, Li Z, Lendlein A, Ma N. Genetic engineering of mesenchymal stem cells by non-viral gene delivery. Clin Hemorheol Microcirc 2014;58(1):19–48.
[74] Mellott AJ, Forrest ML, Detamore MS. Physical non-viral gene delivery methods for tissue engineering. Ann Biomed Eng 2013;41(3):446–68.
[75] Stewart MP, Langer R, Jensen KF. Intracellular delivery by membrane disruption: mechanisms, strategies, and concepts. Chem Rev 2018;118(16):7409–531.
[76] Kim JY, Choi JH, Kim SH, Park H, Lee D, Kim GJ. Efficacy of gene modification in placenta-derived mesenchymal stem cells based on nonviral electroporation. Int J Stem Cells 2021;14(1):112–8.
[77] Escoffre JM, Zeghimi A, Novell A, Bouakaz A. In-vivo gene delivery by sonoporation: recent progress and prospects. Curr Gene Ther 2013;13(1):2–14.
[78] Tomizawa M, Shinozaki F, Motoyoshi Y, Sugiyama T, Yamamoto S, Sueishi M. Sonoporation: gene transfer using ultrasound. World J Methodol 2013;3(4):39–44.
[79] Nishimura K, Ogawa K, Kawaguchi M, Fumoto S, Mukai H, Kawakami S. Suppression of peritoneal fibrosis by sonoporation of hepatocyte growth factor gene-encoding plasmid DNA in mice. Pharmaceutics 2021;13(1).
[80] Chen X, Wan JM, Yu AC. Sonoporation as a cellular stress: induction of morphological repression and developmental delays. Ultrasound Med Biol 2013;39 (6):1075–86.
[81] Zu H, Gao D. Non-viral vectors in gene therapy: recent development, challenges, and prospects. AAPS J 2021;23(4):78.
[82] Shim MS, Wang X, Ragan R, Kwon YJ. Dynamics of nucleic acid/cationic polymer complexation and disassembly under biologically simulated conditions using in situ atomic force microscopy. Microsc Res Tech 2010;73(9):845–56.
[83] Weng Y, Xiao H, Zhang J, Liang XJ, Huang Y. RNAi therapeutic and its innovative biotechnological evolution. Biotechnol Adv 2019;37(5):801–25.
[84] Pezzoli D, Chiesa R, De Nardo L, Candiani G. We still have a long way to go to effectively deliver genes! J Appl Biomater Funct Mater 2012;10(2):82–91.
[85] Jones CH, Chen CK, Ravikrishnan A, Rane S, Pfeifer BA. Overcoming nonviral gene delivery barriers: perspective and future. Mol Pharm 2013;10(11):4082–98.

[86] Knudsen KB, Northeved H, Kumar PE, Permin A, Gjetting T, Andresen TL, et al. In vivo toxicity of cationic micelles and liposomes. Nanomedicine 2015;11(2):467–77.
[87] Barua S, Ramos J, Potta T, Taylor D, Huang HC, Montanez G, et al. Discovery of cationic polymers for non-viral gene delivery using combinatorial approaches. Comb Chem High Throughput Screen 2011;14(10):908–24.
[88] Wahane A, Waghmode A, Kapphahn A, Dhuri K, Gupta A, Bahal R. Role of lipid-based and polymer-based non-viral vectors in nucleic acid delivery for next-generation gene therapy. Molecules 2020;25(12).
[89] Hu H, Wang H, Liang S, Li X, Wang D. Synthesis and characterization of a PAMAM dendrimer nanocarrier functionalized by HA for targeted gene delivery systems and evaluation in vitro. J Biomater Sci Polym Ed 2021;32(2):205–28.
[90] Kang Z, Meng Q, Liu K. Peptide-based gene delivery vectors. J Mater Chem B 2019;7(11):1824–41.
[91] Chen J, Guan X, Hu Y, Tian H, Chen X. Peptide-based and polypeptide-based gene delivery systems. Top Curr Chem 2017;375(2):32.
[92] Baoum A, Ovcharenko D, Berkland C. Calcium condensed cell penetrating peptide complexes offer highly efficient, low toxicity gene silencing. Int J Pharm 2012;427(1):134–42.
[93] Vicinanza C, Lombardi E, Da Ros F, Marangon M, Durante C, Mazzucato M, et al. Modified mesenchymal stem cells in cancer therapy: a smart weapon requiring upgrades for wider clinical applications. World J Stem Cells 2022;14(1):54–75.
[94] Qi C, Jin Y, Chen Y, Li W, Li Y, Liang K, et al. TGase-mediated cell membrane modification and targeted cell delivery to inflammatory endothelium. Biomaterials 2021;269, 120276.
[95] Wu HH, Zhou Y, Tabata Y, Gao JQ. Mesenchymal stem cell-based drug delivery strategy: from cells to biomimetic. J Control Release 2019;294:102–13.
[96] Pascucci L, Coccè V, Bonomi A, Ami D, Ceccarelli P, Ciusani E, et al. Paclitaxel is incorporated by mesenchymal stromal cells and released in exosomes that inhibit in vitro tumor growth: a new approach for drug delivery. J Control Release 2014;192:262–70.
[97] Hassanzadeh A, Altajer AH, Rahman HS, Saleh MM, Bokov DO, Abdelbasset WK, et al. Mesenchymal stem/stromal cell-based delivery: a rapidly evolving strategy for cancer therapy. Front Cell Dev Biol 2021;9, 686453.
[98] Wang X, Gao J, Ouyang X, Wang J, Sun X, Lv Y. Mesenchymal stem cells loaded with paclitaxel-poly(lactic-co-glycolic acid) nanoparticles for glioma-targeting therapy. Int J Nanomedicine 2018;13:5231–48.
[99] Takayama Y, Kusamori K, Tsukimori C, Shimizu Y, Hayashi M, Kiyama I, et al. Anticancer drug-loaded mesenchymal stem cells for targeted cancer therapy. J Control Release 2021;329:1090–101.
[100] Dimitrov DS. Therapeutic proteins. Methods Mol Biol 2012;899:1–26.
[101] Thapa B, Kc R, Uludağ H. TRAIL therapy and prospective developments for cancer treatment. J Control Release 2020;326:335–49.
[102] Mueller LP, Luetzkendorf J, Widder M, Nerger K, Caysa H, Mueller T. TRAIL-transduced multipotent mesenchymal stromal cells (TRAIL-MSC) overcome TRAIL resistance in selected CRC cell lines in vitro and in vivo. Cancer Gene Ther 2011;18(4):229–39.
[103] Park SA, Han HR, Ahn S, Ryu CH, Jeun SS. Combination treatment with VPA and MSCs-TRAIL could increase anti-tumor effects against intracranial glioma. Oncol Rep 2021;45(3):869–78.

[104] Mohme M, Maire CL, Geumann U, Schliffke S, Dührsen L, Fita K, et al. Local intracerebral immunomodulation using interleukin-expressing mesenchymal stem cells in glioblastoma. Clin Cancer Res 2020;26(11):2626–39.

[105] Mahasa KJ, de Pillis L, Ouifki R, Eladdadi A, Maini P, Yoon AR, et al. Mesenchymal stem cells used as carrier cells of oncolytic adenovirus results in enhanced oncolytic virotherapy. Sci Rep 2020;10(1):425.

[106] Ahmed AU, Rolle CE, Tyler MA, Han Y, Sengupta S, Wainwright DA, et al. Bone marrow mesenchymal stem cells loaded with an oncolytic adenovirus suppress the anti-adenoviral immune response in the cotton rat model. Mol Ther 2010;18(10):1846–56.

[107] Li Y, Huang P, Peng H, Yue H, Wu M, Liu S, et al. Antitumor effects of Endostar(rh-endostatin) combined with gemcitabine in different administration sequences to treat Lewis lung carcinoma. Cancer Manag Res 2019;11:3469–79.

[108] Wei X, Liu L, Wang G, Li W, Xu K, Qi H, et al. Potent antitumor activity of the Ad5/11 chimeric oncolytic adenovirus combined with interleukin-24 for acute myeloid leukemia via induction of apoptosis. Oncol Rep 2015;33(1):111–8.

[109] Yang M, Yang C, Tao Y, Tang J, Huang Q, Guo W, et al. Combination therapy with F5/35 fiber chimeric conditionally replicative adenoviruses expressing IL-24 enhances the antitumor effect of temozolomide against melanoma. Cancer Med 2018;7(12):5928–42.

[110] Zhang K, Wang Y, Yu X, Shi Y, Yao Y, Wei X, et al. Recombinant human endostatin combined with radiotherapy inhibits colorectal cancer growth. BMC Cancer 2017;17(1):899.

[111] Zhang J, Chen H, Chen C, Liu H, He Y, Zhao J, et al. Systemic administration of mesenchymal stem cells loaded with a novel oncolytic adenovirus carrying IL-24/endostatin enhances glioma therapy. Cancer Lett 2021;509:26–38.

[112] Muhammad T, Sakhawat A, Khan AA, Ma L, Gjerset RA, Huang Y. Mesenchymal stem cell-mediated delivery of therapeutic adenoviral vectors to prostate cancer. Stem Cell Res Ther 2019;10(1):190.

Index

Note: Page numbers followed by *f* indicate figures and *t* indicate tables.

L

M

N

O

P

R

9780323953467